NONPARAMETRIC METHODS IN COMMUNICATIONS

ELECTRICAL ENGINEERING AND ELECTRONICS

A Series of Reference Books and Textbooks

Editors

Marlin O. Thurston
Department of Electrical Engineering
The Ohio State University
Columbus, Ohio

William Middendorf
Department of Electrical
and Computer Engineering
University of Cincinnati
Cincinnati, Ohio

1. Rational Fault Analysis
   edited by Richard Saeks and S. R. Liberty

2. Nonparametric Methods in Communications
   edited by P. Papantoni-Kazakos and Dimitri Kazakos

Other Volumes in Preparation

NONPARAMETRIC METHODS IN COMMUNICATIONS

Edited by

P. PAPANTONI-KAZAKOS
Department of Electrical Engineering
Rice University
Houston, Texas

DIMITRI KAZAKOS
Department of Electrical Engineering
State University of New York
at Buffalo
Buffalo, New York

MARCEL DEKKER, INC. New York and Basel

Library of Congress Cataloging in Publication Data

Main entry under title:

Nonparametric methods in communications.

(Electrical engineering and electronics ; 2)
1. Statistical communication theory. 2. Nonparametric statistics. I. Papantoni-Kazakos, P., 1945- II. Kazakos, D., 1944- III. Series.
TK5102.5.N568 621.38'01'82 77-14049
ISBN 0-8247-6660-1

MARCEL DEKKER, INC.
270 Madison Avenue, New York, New York 10016

Current printing (last digit):
10 9 8 7 6 5 4 3 2 1

PRINTED IN THE UNITED STATES OF AMERICA

PREFACE

The obsession of optimally designing a system for a particular engineering environment has been prevalent among systems scientists. The complexity of the optimal system naturally increases with the complexity of the assumed model. Intrigued by the mathematics, we tend to generalize our models of the environment and the danger of overmodeling always lurks. We may not have enough data to identify a complicated model structure. Also, the design may be brittle in the sense that slight variations in the assumed model may cause a serious mismatch to the designed system.

A way to overcome the above difficulties is to acknowledge during the design step that there are many gaps in our knowledge of the engineering environment in question. It is possible to start talking about optimal systems with respect to a family of models instead of a single one, and thus bravely admit our ignorance.

In statistical communication, the knowledge of the environment is frequently summarized by a probability density function. Hence, our knowledge limitations are represented by assuming that the probability density function belongs to a class. Nonparametric methods attempt to design systems that perform well for such situations. The term nonparametric means that the classes in question are not characterized by a number of parameters, hence they are larger than parametric classes. Such a property provides us with greater flexibility in modeling real environments. Hence, nonparametric methods are quite practical for real world problems.

The editors are deeply grateful to Professor Demetrios G. Lainiotis of the Electrical Engineering Department, State University of New York at Buffalo. His inspiration, leadership and help were instrumental in starting the present book.

P. Papantoni-Kazakos
Dimitri Kazakos

CONTRIBUTORS

Saleem A. Kassam, Department of Systems Engineering, Moore School of Electrical Engineering, University of Pennsylvania, Philadelphia, Pennsylvania

Dimitri Kazakos, Department of Electrical Engineering, State University of New York at Buffalo, Buffalo, New York

Ludwik Kurz, Department of Electrical Engineering and Electrophysics, Polytechnic Institute of New York, Brooklyn, New York

J. W. Modestino, Department of Electrical and Systems Engineering, Rensselaer Polytechnic Institute, Troy, New York

P. Papantoni-Kazakos, Department of Electrical Engineering, Rice University, Houston, Texas

John B. Thomas, Department of Electrical Engineering, Princeton University, Princeton, New Jersey

CONTENTS

Chapter 1

INTRODUCTION AND BACKGROUND

John B. Thomas

Department of Electrical Engineering
Princeton University
Princeton, New Jersey

I. PRELIMINARIES

During the past twenty-five years, a very large engineering literature has developed on the detection of signals in noise (detection theory) and on the estimation of various parameters of these signals (estimation theory). In most of the literature on detection theory, detectors are sought which are optimal in the Neyman-Pearson sense or which minimize the Bayes risk. These are usually parametric detectors; that is, detectors designed to be optimal for testing between parametric hypotheses where an essentially complete statistical description of each hypothesis is available.

The commonest problem considered is that of choosing between the two hypotheses H_0 and H_1 where

$$H_0 : x(t) = n(t)$$

and $$0 \leq t \leq T \tag{1}$$

$$H_1 : x(t) = n(t) + S(t)$$

Here $x(t)$ is the observation from which the choice is to be made, $n(t)$ is a sample function of a noise process $N(t)$, and $S(t)$ is a known signal although it may contain unknown or random parameters. This class of problems has been treated in great detail and a number of excellent texts exist [1-4] as well as a voluminous literature.

Both the Neyman-Pearson and Bayes approach to optimal detector design require the formation of the likelihood ratio $L(\bar{x})$ and its comparison to a threshold C where

$$L(\bar{x}) = p_1(\bar{x})/p_0(\bar{x}) \gtrless C \tag{2}$$

Here $p_i(\bar{x})$ is a joint density function describing the observation process $X(t)$ under H_i for $i = 0,1$. The likelihood ratio $L(\bar{x})$ [or its logarithm $\log L(\bar{x})$] is then interpreted as a system operating on the observation $x(t)$, yielding the optimal detector.

The principal difficulty that arises in optimal detector design is the need for a tractable expression for the joint densities $p_i(\bar{x})$ for $i = 0,1$. The result, in general, has been the severe restriction that the noise process $N(t)$ be Gaussian. Then $N(t)$ is represented by the Karhunen-Loeve expansion [5] as

$$N(t) = \sum_{i=1}^{\infty} N_i \varphi_i(t) \qquad 0 \leq t \leq T \tag{3}$$

The sequence $\{\varphi_i(t)\}$ is a complete orthonormal set of functions generated by the homogeneous integral equation

$$\varphi(t) = \lambda \int_0^T R(t,s)\varphi(s)\, ds \qquad 0 \leq t \leq T \tag{4}$$

where $R(t,s)$ is the autocorrelation function of the noise process; that is,

$$R(t,s) = E\{X(t)\, X(s)\} \tag{5}$$

The sequence $\{N_i\}$ is a sequence of uncorrelated random variables and the right side of Eq. (3) converges mean-square to the noise process $N(t)$. Each coefficient N_i is given by the linear operation

$$N_i = \int_0^T N(t)\, \varphi_i(t)\, dt \tag{6}$$

If $N(t)$ is a Gaussian random process, then N_i is a normal random variable and the uncorrelated sequence $\{N_i\}$ becomes an independent sequence. The set $\{N_i\}$ are now used to represent $N(t)$ and their joint density (of a particularly simple form because of their independence) is used for $p_0(\bar{x})$. For details, see References 1, 2, or 3.

Unfortunately, when $N(t)$ is not Gaussian, little can be said about the N_i except that they are uncorrelated. This property by itself is not strong enough to allow tractable expressions for the joint densities of the N_i and the Karhunen-Loeve expansion has proved generally useful in optimal detector design only for the Gaussian noise case. For continuous-time observations, little progress seems to have been made along other lines to overcome this fundamental problem. The overall result has been that most continuous-time detection problems assume an additive Gaussian noise [2,3].

In discrete time, the situation is considerably simpler -- at least in principle. Now Eq. (1) is reformulated as

$$
\begin{aligned}
&H_0 : x_i = n_i \\
\text{and} \qquad & \qquad\qquad\qquad i = 1,2,\ldots,N \qquad\qquad (7)\\
&H_1 : x_i = n_i + S_i
\end{aligned}
$$

If the $\{X_i\}$ are independent, then the joint densities $p_0(\bar{x})$ each become the product of N univariate densities and the optimal detection problem is tractable for almost any noise distribution [6]. Further simplification results, of course, if it can be assumed in addition that the $\{N_i\}$ are identically distributed.

As a practical matter, two difficulties still persist with the situation described by Eq. (7). The first is that the assumption of independent observations may not be justified. This is no problem if the $\{N_i\}$ are Gaussian, but if they are not, then the same kind of problems arise with joint densities as in the continuous-time case. If Eq. (7) arises from the sampling of continuous-time processes, then it may not be practical or possible to sample slowly enough to obtain independence. The second difficulty exists with both Eq. (1) and Eq. (7). It is the fact that the statistics of the noise process $\{N(t), 0 \leq t \leq T\}$ may not be known exactly. If such is the case, then the likelihood ratio $L(\bar{X})$ cannot be formed since neither $p_0(\bar{x})$ or $p_1(\bar{x})$ are known.

One way to circumvent a lack of exact knowledge of the noise statistics is to assume much more general hypotheses H_0 and H_1. For simplicity, use the notation of Eq. (7) and assume that the noise observations $\{n_i\}$ are realizations of a sequence $\{N_i\}$ of independent, identically distributed (i.i.d.) random variables. Let $f(x)$ and $F(x)$ be, respectively, the common density function and the common cumulative distribution function for each of the N_i. The $\{S_i\}$ are presumed to be known. For the case where the noise statistics are known, Eq. (7) is equivalent to

$$H_0 : x_i \text{ has distribution } F(x)$$

and $\qquad i = 1,2,\ldots,N \qquad (8)$

$$H_1 : x_i \text{ has distribution } F(x - S_i)$$

If the noise statistics are unknown (or, at least, not known exactly), we might consider generalizing Eq. (8) to the nonparametric hypotheses

$$H_0 : F(0) = 1/2$$

and $\qquad i = 1,2,\ldots,N \qquad (9)$

$$H_1 : F(S_i) = 1/2$$

Now we are assuming only that the noise distribution has zero median and are testing for a shift in median due to the signal S_i . If we can assume further that the noise density is symmetric, then Eq. (9) becomes the more restricted but still nonparametric

$$H_0 : f(x) = f(-x)$$

and $\qquad i = 1,2,\ldots,N \qquad (10)$

$$H_1 : f(x - S_i) = f(S_i - x)$$

The detection problems given by Eqs. (9) and (10) are really very general insofar as the noise statistics are concerned, but they are typical of the kinds of nonparametric hypotheses that are tractable. The goal of nonparametric detection theory is to find optimal or quasi-optimal detectors for such nonparametric hypothesis-testing situations. It turns out that many of these detectors perform surprisingly well when compared to detectors designed to be optimal for much smaller subclasses of hypotheses where, however, much more is assumed to be known about the noise statistics. Furthermore, the nonparametric detectors may perform much better than a detector designed to be optimal for a given parametric hypothesis set, when the assumptions made about the noise statistics turn out to be incorrect. Thus the nonparametric detectors often have the two advantages of 1) affording reasonably good performance compared to their parametric counterparts and 2) offering considerable protection against the consequences of ignorance about the noise statistics. The principal disadvantages are 1) some (but not all) such detectors are difficult to implement and 2) they are usually too conservative.

The literature of nonparametric statistics is very large. A bibliography [7] published in 1962 gives more than 3000 references and a three-volume handbook [8] is required to list the principal nonparametric procedures available in 1968. Furthermore, a number of elementary textbooks devoted entirely to nonparametric statistics have appeared [9-12], as well as some more advanced texts [13]. The engineering literature is growing (see reference [14] for an elementary survey in 1970) and at least one interesting monograph [15] has appeared. One of the purposes of this book is to provide a more up-to-date and more comprehensive bibliography of the engineering (and statistical) literature on nonparametric techniques in signal detection.

II. NOTATION AND TERMINOLOGY

As already discussed in Section I, the observations or input data will be denoted by $\bar{x}$ and will usually be the finite sequence $\{x_i, i = 1,2,\ldots,N\}$, assumed to be a realization of a random sequence $\bar{X}$. The density function of $\bar{X}$ will be denoted by $f(\bar{x}) = P$ and will comprise a complete statistical description of $\bar{x}$. It will be assumed either (1) that $\bar{x}$ consists of "noise only" in which case it will be said that the <u>hypothesis</u> or <u>null hypothesis</u> H_0 is in effect or (2) that $\bar{x}$ consists of "noise and signal" in which case it will be said that the <u>alternative</u> or <u>alternative hypothesis</u> H_1 is in effect. An hypothesis (or alternative) will be called <u>simple</u> if it consists of one (and only one) completely specified form for P; otherwise it will be said to be composite. A composite hypothesis (or alternative) will be called nonparametric if it cannot be described by a finite number of real parameters.

A detector D is a device or system which examines $\bar{x}$ and decides that either the hypothesis* H_0 or the alternative* H_1 is in effect, corresponding to "noise only" or "signal and noise," respectively. Thus, the detector D is a system whose input is the data set $\bar{x}$ and whose output is the decision function (or statistical test function) $D(\bar{x})$. This function is a random variable with the possible values 0 and 1 ; if $D(\bar{x}) = 0$, we accept H_0 ; if $D(\bar{x}) = 1$, we accept H_1 .

The decision function $D(\bar{X})$ is a random variable with possible values 0 and 1. The mathematical expectation of $D(\bar{X})$ will be denoted by $\beta_D(P)$ where

$$\beta_D(P) = E\{D(\bar{X})\} \tag{11}$$

and will satisfy the inequality

$$0 \leq \beta_D(P) \leq 1 \tag{12}$$

The quantity $\beta_D(P)$ is the power function or operating characteristic of the detector D and is the probability of acceptance of H_1 (or rejection of H_0). It is clear that it depends on the form of the detector and on the distribution P of the input data set $\bar{x}$. In this treatment, we accept the viewpoint of the Neyman-Pearson criterion [16] that $\beta_D(P)$ describes the detector performance. Two detectors will be said to be equivalent if they have the same operating characteristic.

---

* For brevity, these two situations will be denoted by "$P \in H_0$" and "$P \in H_1$" , respectively. This notation is inconsistent (but conventional) and since P is a mathematical function while H_0 and H_1 are hypotheses.

It is clear that the detector D can commit two kinds of errors: Type I errors where H_0 is falsely rejected and Type II errors where H_1 is falsely rejected. For obvious reasons, errors of Type I are called false alarms and errors of Type II are called misses. If $P \in H_0$, then $\beta_D(P)$ is the false alarm probability while the quantity

$$\sup_{P \in H_0} \beta_D(P) = \alpha \tag{13}$$

is the size, significance level, or maximum false alarm rate of the detector D. If $P \in H_1$ then $\beta_D(P)$ is the probability of (signal) detection or power of the detector D for the alternative P. The object of most detector design is to make the false alarm probability as small as possible while making the probability of detection as large as possible subject to constraints on realizability, cost, complexity, etc.

Let Δ be the class of all detectors satisfying

$$\beta_D(P) \leq \alpha \qquad \text{for all} \quad P \in H_0 \tag{14}$$

A detector $D^* \in \Delta$ is said to be optimum, in the Neyman-Pearson sense (for hypothesis H_0, alternative H_1, and level α), if

$$B_{D^*}(P) \geq \beta_D(P) \qquad \text{for all} \quad P \in H_1 \quad \text{and all} \quad D \in \Delta : \tag{15}$$

that is, if D^* simultaneously maximizes all detection probabilities of interest subject to the condition that its false alarm rate be no greater than the desired level α.

If H_0 and H_1 are simple, an optimum detector can always be constructed according to the procedure given by the Neyman-Pearson fundamental lemma of hypothesis testing [12]. If H_0 and H_1 are composite, an optimum detector usually does not exist. In such cases, one may search for a detector which is optimum within some class of detectors of special form. For example, a detector is said to be unbiased [16] (at level α) if Eq. (14) holds and if, in addition, $\beta_D(P) \geq \alpha$ for all $P \in H_1$; a detector D^* is said to be optimum unbiased if Eq. (15) holds when Δ is the class of unbiased detectors. One may also seek a detector which has an appropriate local optimum property [16], in the sense that Eq. (15) is satisfied not necessarily for all $P \in H_1$, but only for certain weak signal conditions; i.e., only in the limit as P becomes "close" to H_0 in a suitably restricted sense. Finally, one may consider some reasonable preliminary reduction of the input data and then seek useful detectors among those which employ only the reduced data. Data reduction may be justified by invariance considerations [16], or it may be the result of such operations as frequency conversion, filtering, clipping, sampling of continuous waveforms, and ranking or grouping of discrete observations. In particular, all detectors discussed here, unless otherwise stated, base their decision upon a finite set $(x_1,x_2,\ldots,x_N)$ of independent samples of the input process. This assumption constitutes a strong restriction on the input, but it will often be reasonable to suppose that sufficiently many such samples can be obtained from the processes of practical interest, if the total observation time is large in comparison to the reciprocal of the bandwidth.

Let $x_1,x_2,\ldots,x_N$ be independent observations of the input, and let $D(x_1,x_2,\ldots,x_N)$ be a detector based on the sample data; D is said to be consistent [16] for an alternative H_1 if for any $P \in H_1$,

$$\lim_{N\to\infty} \beta_{D(x_1,x_2,\ldots,x_N)}(P) = 1 \tag{16}$$

that is, if the probability of detection for any fixed alternative can be made as large as desired when sufficiently many observations are available. It is tacitly assumed that for the hypothesis H_0 of interest, D has the same false alarm rate α for each N, or if H_0 is a nonparametric hypothesis, that D is at least asymptotically nonparametric or robust [17] in the sense that for all $P \in H_0$

$$\lim_{N \to \infty} \beta_{D(x_1, x_2, \ldots, x_N)}(P) = \alpha \tag{17}$$

As a general rule, each detection problem has many possible solutions; that is, given an hypothesis H_0 and an alternative H_1, there are many (non-equivalent) detectors which can be implemented. Therefore it is necessary to develop means for comparing the usefulness of various detectors. In this respect, two of the commonest situations encountered is to compare

(1) two nonparametric detectors under the same hypothesis H_0,

or

(2) a nonparametric detector and a parametric detector, both operating under a subclass of the hypothesis H_0 (and/or alternative H_1).

The usual method of comparing two detectors (tests) for the same hypothesis H_0 consists of computing the relative efficiency of one detector with respect to the other. Although it is possible to define many relative efficiencies, the one most commonly used is due to Pitman. Let $N_1 = n_{D_1}(\alpha, \beta, P)$ and $N_2 = n_{D_2}(\alpha, \beta, P)$ be the smallest number of observations required for detectors D_1 and D_2 to attain a probability of detection at least equal to β for the alternative $P \in H_1$ and for the same false alarm probability α under the same hypothesis H_0. The relative efficiency $e_{2,1}$ of detector D_2 with respect to detector D_1 is defined as

$$e_{2,1} = \frac{n_{D_1}(\alpha, \beta, P)}{n_{D_2}(\alpha, \beta, P)} \tag{18}$$

It should be clear that $e_{2,1}$ is a function of α, H_1, N_1, and N_2; and it may be difficult to evaluate for arbitrary α, H_1, N_1, and N_2.

It is often simplest to let N_1 and N_2 approach infinity. Since, in such a case, the probability of detection (power) will approach unity (for consistent tests against fixed alternatives) it is necessary to allow the alternative H_1 to approach the hypothesis H_0 in such a way that the probability of detection (power) β remains constant as N_1 and $N_2 \to \infty$. We are lead thus to the concept of asymptotic relative efficiency (ARE), $E_{1,2}$ where

$$E_{1,2} = \lim_{\substack{N_1 \to \infty \\ N_2 \to \infty \\ H_1 \to H_0}} e_{1,2}(\alpha_1, H_1, N_1, N_2)\Big|_{\beta=\beta_1} \tag{19}$$

and α_1 and β_1 are fixed values of α and β. In many cases, $E_{1,2}$ is just a number, independent of α.

Since the definition of ARE has been obtained by allowing H_1 to approach H_0, it is basically a weak-signal measure of detector performance. It is clear that it may provide a useful figure of merit less than unity for a detector D_2 compared to a detector D_1 which is optimal for the (H_0, H_1) being considered. On the other hand, the ARE of D_2 compared to D_1 may exceed unity for some (H_0, H_1) for which D_1 is not optimal; that is, it may be possible to obtain superior performance from a suboptimal detector which is less sensitive to departures from assumptions and which is operating in an environment different from that for which the optimal detector was designed.

In the subsequent sections, we will discuss and compare some of the simplest and commonest parametric and nonparametric detection schemes. The most disconcerting aspect of nonparametric detection is the lack of a simple constructive procedure corresponding to the Neyman-Pearson lemma for the parametric case. There would appear to be no way around this problem since it is an inherent aspect of nonparametric techniques. As mentioned earlier, however, elementary nonparametric statistics is relatively well-organized. Also, it might be well to keep in mind that the structural simplicity and compactness of parametric detection theory is somewhat misleading. As mentioned before the only parametric problem that has received extensive attention is the rather specialized one where the input under the hypothesis is stationary Gaussian noise.

III. ONE-INPUT DETECTORS

In this section we consider the case where the input data is a single sequence

$$\bar{x} = (x_1, x_2, \ldots, x_N) \tag{20}$$

of independent random variables each with density function $f(x_i)$ and cumulative distribution function $F_i(x) = P\{X_i \leq x\}$. The mean of each X_i will be denoted by μ_i and the variance by σ_i^2.

The one-input problem most often treated [2] is the detection of a sure (deterministic) signal $\bar{S} = \{S_1, S_2, \ldots, S_N\}$ in white Gaussian noise $\{n_1, n_2, \ldots, n_N\}$ of zero mean and fixed variance σ_0^2. In this case, the simple hypothesis H_0 corresponds to $x_i = n_i$ and the composite (parametric) alternative H_1 corresponds to $x_i = S_i + n_i$ so that

$$H_0 : F_i(x) \quad \text{is Gaussian}, \; \mu_i = 0, \; \sigma_i^2 = \sigma_0^2$$

and

$$H_1 : F_i(x) \quad \text{is Gaussian}, \; \mu_i = S_i, \; \sigma_i^2 = \sigma_o^2 \tag{21}$$

The joint density functions of the x_i under the hypothesis H_0 and the alternative H_1 will be denoted by $f_0(\bar{x})$ and $f_1(\bar{x})$, respectively, where it is easy to see that [2]

$$f_0(\bar{x}) = (2\pi\sigma_0^2)^{-n/2} \exp[-\sum_{k=1}^{n} x_k^2/2\sigma_0^2] \tag{22}$$

and

$$f_1(\overline{x}) = (2\pi\sigma_0^2)^{-n/2} \exp[-\sum_{k=1}^{n} (x_k - S_k)^2/2\sigma_0^2] \tag{23}$$

According to the Neyman-Pearson lemma [16], the likelihood ratio

$$L(\overline{x}) = f_1(\overline{x})/f_0(\overline{x}) \tag{24}$$

is calculated for the input data $\overline{x}$ and is compared to a threshold C_0. If $L(x) < C_0$, then $D(\overline{x}) = 0$ and we accept H_0, "noise only". If $L(\overline{x}) > C_0$, then $D(\overline{x}) = 1$ and we accept H_1, "signal and noise."

In this particular case, the likelihood ratio is easy to find by inspection and is

$$L(\overline{x}) = \exp[-\sum_{k=1}^{n} (2S_k x_k - x_k^2)/2\sigma_0^2] \tag{25}$$

Since σ_0^2 and each S_k are known, it is clear that comparing $L(\overline{x})$ to a threshold C_0 is equivalent to comparing

$$\sum_{k=1}^{N} S_k x_k$$

to some other number C_0' so that the detector D_0 becomes

$$\sum_{k=1}^{N} S_k x_k \begin{cases} < C_0' \Rightarrow D_0(\overline{x}) = 0 \\ \\ > C_0' \Rightarrow D_0(\overline{x}) = 1 \end{cases} \tag{26}$$

Here C_0' (or C_0) are obtained by fixing the fixing the false alarm probability α, as discussed in detail in [2].

The detector D_0 of Eq. (26) is often called a <u>correlation detector</u> or <u>correlator</u> since it calculated the time cross correlation function of the (known) signal $\overline{S}$ and the input data $\overline{x}$. The correlation detector is implemented by a <u>matched filter</u>; thus the left side of Eq. (26) can be obtained by passing the input data sequence $\overline{x} = (x_1, x_2, \ldots, x_N)$ through a filter with impulse response h_k given by

$$h_k = \begin{cases} S_{n-k} & , k = 1,2,\ldots,n \\ 0 & , \text{elsewhere} \end{cases} \tag{27}$$

A simpler case occurs when the signal is taken to be an unknown positive constant, corresponding to the detection of a positive d-c signal of unknown amplitude in Gaussian noise. Now the simple hypothesis H_0 and the composite (parametric) alternative H_1 are

$$H_0 : F_i(x) \quad \text{is Gaussian}, \ \mu_i = 0, \ \sigma_i^2 = \sigma_0^2$$

and (28)

$$H_1 : F_i(x) \quad \text{is Gaussian}, \ \mu_i = \mu, \ \sigma_i^2 = \sigma_0^2$$

Since S_k is constant for all k, the detector D_0 of Eq. (26) becomes the <u>linear detector</u> D_1 given by

$$\sum_{k=1}^{N} x_k \begin{cases} < C_1' \Rightarrow D_1(x) = 0 \\ > C_1' \Rightarrow D_1(x) = 1 \end{cases} \tag{29}$$

where, as before, C_1' is obtained by fixing α. In this case, it is easy to see that the density function of

$$\sum_{k=1}^{N} x_k$$

under H_0 is a Gaussian density with zero mean and variance $N\sigma_0^2$. Thus C_1' is obtained from

$$\alpha = \frac{1}{\sqrt{2\pi N}\,\sigma_0} \int_{C_1'}^{\infty} e^{-x^2/2N\sigma_0^2} \, dx \tag{30}$$

or, equivalently,

$$C_1' = \sqrt{N}\,\sigma_0 K \tag{31}$$

where K is found from

$$\Phi(k) = 1 - \alpha \tag{32}$$

and $\Phi(\cdot)$ is the unit normal cumulative distribution function.

Let us now consider a much broader class of detection problems than those described by Eqs. (21) and (28). In the process of so doing, it will become clear that the median of a distribution is a natural generalization in the nonparametric case of the mean of a Gaussian distribution.

Let the x_i of Eq. (20) be independent and identically distributed with the common cumulative distribution function $F(x) = P\{x_i \le x\}$. Define the probability p by

$$p = P\{x_i > 0\} = 1 - F(0) \tag{33}$$

Consider the <u>nonparametric hypothesis</u> H_0 and the <u>nonparametric alternative</u> H_1 where

$$H_0 : p = 1/2 \quad , \quad F(x) \text{ otherwise arbitrary}$$

and (34)

$$H_1 : p > 1/2 \quad , \quad F(x) \text{ otherwise arbitrary}$$

This describes the situation, for example, where under the hypothesis the x_i are independent noise observations with zero medians versus the alternative that the medians of the independent observations are greater than zero due to the presence of an additive positive signal.

A Neyman-Pearson optimal detector $D_2(\bar{x})$ exists for Eq. (34) and is easily found. Let f be the density function of x_i under any signal-present condition (i.e., under any member of H_1). Define f^+ and f^- as the conditional densities of x_i given $x_i > 0$ and $x_i \le 0$, respectively. From the definition of conditional density, it follows that

$$f = pf^+ + (1 - p)f^- \tag{35}$$

Define a new density function f_0 by

$$f_0 = (f^+ + f^-)/2 \tag{36}$$

It is clear that f_0 is a density function and that it belongs to a random variable with zero median since it can be obtained from Eq. (35) by setting $p = 1/2$. Thus f_0 is an acceptable density under the signal-absent condition (i.e., under H_0).

The Neyman-Pearson lemma for testing the simple hypothesis f_0 versus the simple alternative f requires the formation of the likelihood ratio

$$L_2(\bar{x}) = \prod_{i=1}^{N} f(x_i)/f_0(x_i) \tag{37}$$

The multiplicand in this last expression takes on one value if $x_i > 0$ and another if $x_i \leq 0$; that is,

$$\frac{f(x_i)}{f_0(x_i)} = \frac{pf^+}{\frac{1}{2}f^+} = 2p \quad , \quad x_i > 0 \tag{38}$$

and

$$\frac{f(x_i)}{f_0(x_i)} = \frac{(1-p)f^-}{\frac{1}{2}f^-} = 2(1-p) \quad , \quad x_i \leq 0 \tag{39}$$

Thus Eq. (37) becomes

$$L_2(\bar{x}) = 2^N p^{\sum_{i=1}^{N} u(x_i)} (1-p)^{N - \sum_{i=1}^{N} u(x_i)} \tag{40}$$

where $u(x_i)$ is the unit step function given by

$$u(x_i) = \begin{cases} 1 & , \quad x_i > 0 \\ 0 & , \quad x_i \leq 0 \end{cases} \tag{41}$$

It is clear that comparing L_2 to a threshold C_{20} is equivalent to comparing

$$\sum_{i=1}^{n} u(x_i)$$

to another constant C_2'. Thus the Neyman-Pearson detector $D_2(\bar{x})$ is the so-called <u>sign detector</u> and is given by

$$\sum_{i=1}^{N} u(x_i) \begin{cases} < C_2' \Rightarrow D_2(\bar{x}) = 0 \\ \\ > C_2' \Rightarrow D_2(x) = 1 \end{cases} \tag{42}$$

As usual, the constant C_2' is determined by fixing the false-alarm rate α. The quantity $\sum_i u(x_i)$ is the number k of positive observations (successes) and is a random variable which is binomially distributed with parameter p. Thus k has a (discrete) density function under H_1 given by the binomial density

$$b(k;N,p) = \binom{N}{k} p^k (1 - p)^{N-k} \quad , \quad k = 0,1,\ldots,N \tag{43}$$

and under H_0 by

$$b(k;N,\tfrac{1}{2}) = \binom{N}{k} (\tfrac{1}{2})^N \quad , \quad k = 0,1,\ldots,N \tag{44}$$

where $\binom{N}{k}$ is the binomial coefficient. The operating characteristic of the detector D_2 is given by Eq. (11) and is

$$\beta_{D_2}(F) = \sum_{k=C_2'+1}^{N} \binom{N}{k} p^k (1 - p)^{N-k} \tag{45}$$

where C_2' is the largest integer satisfying

$$\alpha \leq \sum_{k=C_2'+1}^{N} \binom{N}{k} (\tfrac{1}{2})^N \tag{46}$$

and α is the predetermined false alarm rate.

Note from Eq. (46) that the value of C_2' (and the false alarm rate) is the same for all distributions under H_0; that is, for all input noises with zero medians. Thus the sign detector is a nonparametric detector for H_0. Of more importance, it is clear from Eq. (46) that the sign detector is nonparametric for any hypothesis which is a subclass of that H_0; for example, for the hypotheses of Eqs. (21) and (28). Of course, the sign detector is suboptimal for the testing situation of Eqs. (21) and (28) as well as for any other case where H_0 is a subclass of the H_0 of Eq. (34). However, it is easy to implement; it is nonparametric under very general conditions; and it will turn out that its asymptotic performance (as the number of input data sample points becomes large) is not too poor compared to the performance of optimal detectors. Also, as mentioned previously, the performance of the sign detector may be superior to that of an optimal detector operating under an inappropriate (H_0,H_1).

We proceed now to compare the sign detector of Eq. (42) to the linear detector of Eq. (29) using the concept of ARE given by Eq. (19). We shall evoke a simple form of the Central Limit Theorem which states that sums of independent and identically distributed random variables tend in the limit to normal or Gaussian distributions. In particular, as N becomes large, the binomial density function $b(k;N,p)$ of Eq. (43) with mean Np and variance $Np(1 - p)$ tends to the Gaussian density function

$$\eta(x) = \frac{1}{\sqrt{2\pi}\sigma} \exp[-\tfrac{1}{2}(\frac{x - \mu}{\sigma})^2] \tag{47}$$

with mean $\mu = Np$ and variance $\sigma^2 = Np(1 - p)$. Actually, as pointed out by Feller [18], the convergence to the normal density is very rapid if p is not too far from one-half.

As before, let

$$k = \sum_{i=1}^{N} u(x_i)$$

be the number of positive observations from the input data set $\bar{x}$. In the limit as $N \to \infty$, Eq. (45) becomes the symptotic probability of signal detection β_2 for the sign detector. On evoking the Central Limit Theorem and using Eq. (47), we have

$$\beta_2 = \lim_{N\to\infty} \beta_{D_2}(F) = \int_{C_2'}^{\infty} \eta(x)dx \tag{48}$$

After a change in variable, this last integral can be written as

$$\beta_2 = 1 - \Phi\left[\frac{C_2' - Np}{\sqrt{Np(1 - p)}}\right] = \Phi\left[\frac{Np - C_2'}{\sqrt{Np(1 - p)}}\right] \tag{49}$$

where, as before, $\Phi(\cdot)$ is the unit normal cumulative distribution function. It is clear that the limiting expression for Eq. (46) is obtained by setting $p = \frac{1}{2}$ in Eq. (49); thus the asymptotic false alarm rate for the sign detector is

$$\alpha_2 = \lim_{N\to\infty} \alpha = 1 - \Phi\left[\frac{2C_2' - N}{\sqrt{N}}\right] \tag{50}$$

Now the threshold C_2' may be eliminated from the last two equations to yield

$$\beta_2 = \Phi\left[\frac{\sqrt{N}(2p - 1) - \Phi^{-1}(1 - \alpha_2)}{2\sqrt{p(1 - p)}}\right] \tag{51}$$

where $\Phi^{-1}(\cdot)$ is the inverse of $\Phi(\cdot)$.

Let us now consider the linear detector of Eq. (29) operating on the input data set $\overline{x} = (x_1, x_2, \ldots, x_M)$ consisting of M observations and obeying Eq. (34). If the X_i are Gaussian (as was assumed in deriving the linear detector from the Neyman-Pearson lemma), then

$$\sum_{i=1}^{M} X_i$$

is Gaussian also. On the other hand, if the X_i are only identically distributed with means μ and variances σ^2, the sum

$$\sum_{i=1}^{M} X_i$$

is asymptotically Gaussian with mean $M\mu$ and variance $M\sigma^2$. Thus the asymptotic probability of signal detection β_1 for the linear detector is, from comparison with Eq. (49),

$$\beta_1 = \Phi\left[\frac{M\mu - C_1}{\sqrt{M}\,\sigma}\right] \tag{52}$$

The asymptotic false alarm probability α_1 can be obtained by assuming the X_i have zero mean under H_0. Now

$$\sum_{i=1}^{m} X_i$$

is asymptotically Gaussian under H_0 with zero mean and variance $M\sigma^2$ and α_1 becomes

$$\alpha_1 = \frac{1}{\sqrt{2\pi M\sigma}} \int_{C_1'}^{\infty} e^{-x^2/2M\sigma^2}\, dx = 1 - \Phi\left[\frac{C_1}{\sqrt{M\sigma}}\right] \tag{53}$$

Again, the threshold C_1' may be eliminated from the last two equations to yield

$$\beta_1 = \Phi\left[\sqrt{M}\,\frac{\mu}{\sigma} - \Phi^{-1}(1 - \alpha_1)\right] \tag{54}$$

An examination of Eq. (19) shows that the asymptotic relative efficiency $E_{2,1}$ of the sign detector with respect to the linear detector may be found by setting $\beta_1 = \beta_2$ and $\alpha_1 = \alpha_2$ and solving for the ratio of M to N. The result is

$$E_{2,1} = \lim_{\substack{N\to\infty \\ M\to\infty}} \frac{M}{N} = \frac{\sigma^2 (p - \frac{1}{2})^2}{\mu^2 p(1 - p)} \tag{55}$$

We have already assumed that $p > \frac{1}{2}$, that the noise distribution f is symmetric, and that μ is small (i.e., that a weak-signal condition holds). Thus p may be approximated by

$$p = \tfrac{1}{2} + \int_0^{\mu} f(x - \mu)dx \approx \tfrac{1}{2} + \mu f(0) \tag{56}$$

Now Eq. (55) becomes

$$E_{2,1} \approx \frac{4\sigma^2 f^2(0)}{1 - 4\mu^2 f^2(0)} \tag{57}$$

In the limit of vanishingly small signals, $\mu \to 0$ and we have

$$E_{2,1} \approx 4\sigma^2 f^2(0) \tag{58}$$

This last expression gives the ARE of the sign detector with respect to the linear detector for the detection of a d-c signal in any noise having a symmetric density function f.

If the noise is Gaussian, the $f(0) = 1/\sqrt{2\pi}\,\sigma$ and the right side of Eq. (58) becomes $2/\pi$. Recall that this corresponds to Eq. (28) for which the linear detector is optimal. Thus the sign detector is only about 64% as efficient as the optimal detector for a d-c signal in Gaussian noise. On the other hand, suppose that the noise has a two-sided exponential distribution so that f is given by

$$f(x) = \frac{\alpha}{2} e^{-\alpha|x|} \quad , \quad -\infty < x < \infty \tag{59}$$

This density function corresponds to a random variable with zero mean and variance $\sigma^2 = 2/\alpha^2$. In this case, Eq. (58) becomes $E_{2,1} = 2$ and the sign detector is twice as efficient as the linear detector. More generally, it is obvious that Eq. (58) will be greater than unity whenever

$$\frac{1}{f(0)} < 2\sigma = 2\left[\int_{-\infty}^{\infty} x^2 f(x)dx\right]^{\frac{1}{2}} \tag{60}$$

(Recall that f is not only symmetric, but has zero mean.) Thus any noise with a symmetric density function which falls off slowly enough will yield an ARE greater than unity in comparing the sign detector to the linear detector.

Since the sign detector was derived as the (Neyman-Pearson) optimal detector for Eq. (34), it is clear that it is a natural test for a shift in median. However, it used relatively little information from the input data set and, specifically, does not take account of the distance of an observation x_i from the origin. Instead of

$$\sum_{i=1}^{N} u(x_i)$$

it might be more reasonable to consider some statistic

$$\sum_{i=1}^{N} \lambda_i u(x_i)$$

where the λ_i are weights reflecting the distance of an observation x_i from the origin. Let us rank the observed data $\overline{x} = (x_1, x_2, \ldots, x_N)$ by ordering the absolute values of the x_i so that

$$|x_{k_1}| < |x_{k_2}| < \ldots < |x_{k_n}| \tag{61}$$

Now consider the detector given by

$$\sum_{i=1}^{n} d_i \begin{cases} < C_2' \Rightarrow D_2'(\overline{x}) = 0 \\ \\ > C_2' \Rightarrow D_2'(\overline{x}) = 1 \end{cases} \tag{62}$$

where d_i is given by

$$d_i = \begin{cases} 0, & x_{k_i} \leq 0 \\ \\ i, & x_{k_i} > 0 \end{cases} \tag{63}$$

The detector $D_2'(\overline{x})$ is called the <u>Wilcoxon detector</u> and is easily compared to the linear detector [10,19,20,21]. Space precludes the calculations, but the ARE of the Wilcoxon detector with respect to the linear detector is given by

$$E_{2',1} = 12\sigma^2 \left[\int_{-\infty}^{\infty} f^2(x)\,dx\right]^2 \tag{64}$$

where, as before, f is the noise density (i.e., the density of one of the x_i under H_0 and σ^2 in the noise variance. For the Gaussian case, $E_{2',1}$ becomes

$$E_{2',1} = \frac{3}{\pi}\left[\frac{1}{\sqrt{2\pi}} \int_{-\infty}^{\infty} e^{-y^2/2}\,dy\right]^2 = \frac{3}{\pi} = 0.955 \tag{65}$$

For the two-sided exponential density of Eq. (3-36), we have

$$E_{2',1} = \frac{3}{2}\left[2\alpha \int_0^\infty e^{-2\alpha x}\, dx \right]^2 = \frac{3}{2} = 1.5 \tag{66}$$

Thus the Wilcoxon detector is more than 95% as efficient as the linear detector for Gaussian noise and, as might be expected, considerably better for noise with a two-sided exponential distribution. A more surprise result [21] is that $E_{2',1}$ given by Eq. (64) cannot be less than a positive lower limit of 0.864 for any f .

The Wilcoxon detector of Eq. (62) is a <u>linear rank detector</u> as is the closely related Mann-Whitney detector [13]. Even more impressive results can be obtained with rank statistics which are not linear [13,20]. The Kolmogorov-Smirnov detector is especially interesting [13] and can be expressed as a maximum of a finite number of linear rank detectors. Also included in this general type of detector are the Renyi and the Cramer-von Mises detectors [13]. The complexity of implementation of such schemes grows rapidly, of course. However, with the increasing speed and memory size of digital computers, such detectors become more practical and merit further attention. Another fruitful approach [22,23] is to rank small blocks $(x_{i+1}, x_{i+2}, \ldots, x_{i+k})$, $k \quad N$, of data and form "mixed statistics" or detectors based on naturally ordered blocks of ranked data. This approach reduces greatly the complexity of implementation and the memory and/or storage requirements of the detector. The analyses of most detectors employing ranking tend to be complicated and to rely critically on the applicability of the Central Limit Theorem.

IV. ORGANIZATION OF THE BOOK

The remainder of this book consists of five chapters dealing with various aspects of nonparametric detection (and estimation). Each chapter has been written by a person active in research in the subject area of that chapter.

Chapter 2 has been contributed by Professor J.W. Modestino of Rensselaer Polytechnic Institute. It treats the general area of adaptive nonparametric detection, principally in the context of non-Gaussian noise and models for such noise. Both ranked and non-ranked procedures are discussed. Most of the treatment relates to one-input models such as those discussed in Section III, but two-input or two-channel problems are also considered. Adaptive algorithms and their convergence are treated and numerical results are given.

Chapter 3 has been written by Professor L. Kurz of the Polytechnic Institute of New York. It treats nonparametric detection based on a partitioning of the observation vector $\bar{x}$ into a finite number of quantiles. As in Chapter 2, emphasis is placed on the performance of the resulting detectors in non-Gaussian noise of an impulsive and dispersive nature. The techniques employed here may be considered as alternatives to the rank methods discussed in Section III of this chapter.

Chapter 4 has been written by Professor S. A. Kassam of the University of Pennsylvania. It introduces the idea of conditional statistical tests and applies it to nonparametric detection. Conditional tests are threshold tests as discussed in Section II, but the threshold itself is not fixed _a priori_, but depends directly on the specific data set being observed. The tests treated are easy to implement and lead to nonparametric detection systems.

Chapter 5 has been written by Professor P. Papantoni-Kazakos of Rice University. The subject matter treated here is the performance of rank tests for small sample sizes (small values of N in the observation vector

$\bar{x} = (x_1, x_2, \ldots, x_N)$). The small-sample problem is difficult to deal with since limit theorems (principally, the Central Limit Theorem) cannot be used. Both analytical and numerical results are given for this case which has received relatively little attention in the statistical literature and almost none in engineering applications.

The final chapter, Chapter 6, has been contributed by Professor D. Kazakos of the State University of New York at Buffalo. It concerns some advances in the nonparametric estimation of probability density functions. Although this topic is not in the mainstream of detection theory itself, it has very serious implications in the implementation of detection systems. The underlying assumption in parametric detection is that the statistics of the observation vector are exactly known. Nonparametric <u>detection</u> techniques were developed to avoid this problem. An alternative approach is to <u>estimate</u> the statistics in an adaptive or learning mode. In this sense, Chapter 6 is most closely related in spirit to Chapter 2.

REFERENCES

1. Middleton, D. (1960) <u>Introduction to Statistical Communication Theory</u>. McGraw-Hill Book Co., Inc., New York, N.Y.
2. Helstrom, C.W. (1968) <u>Statistical Theory of Signal Detection.</u> Pergamon Press, New York, N.Y.
3. Van Trees, H.L. (Part I, 1968; Part 3, 1971) <u>Detection, Estimation, and Modulation Theory.</u> McGraw-Hill Book Co., Inc., New York, N.Y.
4. Whalen, A.D. (1971) <u>Detection of Signals in Noise</u>. Academic Press, New York, N.Y.
5. Reference 3, Part 1, Chapter 3.
6. Miller, J.H. and J.B. Thomas (1972) "Detectors for discrete-time signals in non-Gaussian noise," <u>IEEE Trans. on Inf. Th.</u> <u>IT-18(2)</u>: 241-250.

7. Savage, I.R. (1962) Bibliography of Nonparametric Statistics. Harvard University Press, Cambridge, Mass.

8. Walsh, J.E. (Vol. 1, 1962; Vol. 2, 1965; Vol.3, 1968) Handbook of Nonparametric Statistics. Van Nostrand: Princeton, N.J.

9. Siegel, S. (1956) Nonparametric Statistics for the Behavioral Sciences. McGraw-Hill: New York, N.Y.

10. Fraser, D.A.S. (1957) Nonparametric Methods in Statistics. Wiley: New York, N.Y.

11. Noether, G.E. (1967) Elements of Nonparametric Statistics. Wiley: New York, N.Y.

12. Bradley, J.V. (1968) Distribution-Free Statistical Tests. Prentice-Hall: Englewood Cliffs, N.J.

13. Hajek, J. and Z. Sidak (1967) Theory of Rank Tests. Academic Press: New York, N.Y.

14. Thomas, J.B. (1970) "Nonparametric detection," Proc. IEEE 58(5): 623-631.

15. Gibson, J.D. and J.L. Melsa (1975) Introduction to Nonparametric Detection with Applications. Academic Press: New York, N.Y.

16. Lehmann, E.L. (1959) Testing Statistical Hypotheses. Wiley: New York, N.Y., p. 65.

17. Kendall, M.G. and A. Stuart (1962) The Advanced Theory of Statistics. Hafner: New York, N.Y., vol. 2, p. 240.

18. Feller, W. (1968) An Introduction to Probability Theory and Its Application, Vol. 1. Wiley: New York, N.Y., Chap. 7.

19. Millard, J.B. and L. Kurz (1962) "Nonparametric signal detection -- An application of the Kolmogorov-Smirnov, Cramer-von Mises tests (Abstract only)" IEEE Trans. Inf. Th. IT-12: 275.

20. Carlyle, J.W. and J.B. Thomas (1964) "On nonparametric signal detectors," IEEE Trans. Inf. Th. IT-10: 146-152.

21. Hodges, J.L. and E.L. Lehman (1956) "The efficiency of some non-parametric competitors to the t-test," Ann. Math. Stat. 27: 324-335.

22. Feustel, E.A. and L.D. Davisson (1967) "The asymptotic relative efficiency of mixed statistical tests," IEEE Trans. Inf. Th. IT-13: 247-255.

23. Feustel, E.A. and L.D. Davisson (1968) "On the efficacy of mixed locally-most-powerful one- and two-sample rank tests," IEEE Trans. Inf. Th. IT-14: 776-778.

Chapter 2

ADAPTIVE NONPARAMETRIC DETECTION TECHNIQUES

J. W. Modestino

Electrical and System Engineering Department
Rensselaer Polytechnic Institute
Troy, New York

I. INTRODUCTION

A large and growing body of engineering literature exists concerned with the application of nonparametric statistical hypothesis testing procedures to the detection of weak signals in non-Gaussian background noise. It is characteristic of such applications that a complete statistical description of the prevailing noise environment is unavailable for one reason or another. Typically, the classical nonparametric detector structures are implemented as a threshold test on an appropriately defined test statistic. These test statistics are generally computed as some nonlinear functional of the observations.

In situations where asymptotic performance criteria, such as asymptotic relative efficiency (A.R.E.), are appropriate the classical nonparametric detection procedures have demonstrated fairly robust although generally suboptimum performance. This absence of optimum procedures is a major deficiency of nonparametric statistics in general. Indeed, most nonparametric procedures have evolved in an ad hoc manner and exhibit no real optimal properties. An exception is the concept of a locally most powerful test, although here parametric knowledge of the underlying noise statistics is required in order to implement the corresponding test statistic. It is highly unlikely that this information would be available in any realistic non-Gaussian detection application. Nevertheless, the locally most powerful test proves useful in suggesting a basic detector structure whose paramaters can be adapted to the prevailing noise environment. This is especially important since, in those cases where only an incomplete statistical description of the background noise is available, it is highly likely that these statistics also change with time. Thus it is desirable to continuously monitor the noise environment and periodically update the detector parameters while the basic structure remains fixed.

In the present chapter we describe various adaptive nonparametric detection procedures for the detection of weak signals in statistically undefined noise backgrounds. The detection applications to be considered are of both the single-channel and dual-channel variety; the latter case including the situation where an independent reference channel is available as well as the dual channel test of independence. These detector structures have, for the most part, been suggested from consideration of the corresponding locally most powerful detector. The performance criteria will be taken as the A.R.E. generally measured with respect to the corresponding locally most powerful detector. It is shown that the resulting adaptive detector structures are relatively easy to implement and provide efficient and robust performance for a wide range of underlying noise distributions.

The chapter is organized as follows: In Section II the detection situations and noise models to be employed in the remainder of the chapter are defined. This is followed by a discussion of adaptive coincidence procedures in Section III. A coincidence procedure is any detection strategy based only upon the polarity of the observations. Adaptive procedures based upon the relative ranks of the observations are discussed in Section IV. A discussion of selected, although fairly general, adaptive algorithms for updating the various detector parameters is provided in Section V. Finally, a summary and discussion of future research directions can be found in Section VI.

II. PRELIMINARIES

In what follows, the discussion will be in terms of discrete data sequences and filtering operations. Since the signal processing in an increasing number of real-world systems is or will be done digitally, the discrete case is quite appropriate. The noise samples will be assumed independent and identically distributed (i.i.d.). This assumption of sample-to-sample independence, although considerably simplifying the analysis, is admittedly unrealistic in a number of practical applications. Nevertheless, it is expected that sample-to-sample dependence will not result in significant qualitative differences.

In the case of a single-channel detector the received data will be represented as an N-vector, more specifically a column vector, $\underline{y}_N$ given by

$$\underline{y}_N = \theta \underline{s}_N + \underline{n}_N \tag{1}$$

with $\underline{s}_N$ an a priori known signal sequence, $\theta \geq 0$ a scale constant, and $\underline{n}_N$ an additive noise sequence assumed independent of the signal. The noise samples are assumed i.i.d. with common univariate probability density function (p.d.f.) $f(\cdot)$, continuous and symmetric about the original but otherwise arbitrary. In the case of a dual-channel detector with a reference noise channel the observables are described by

$$\underline{x}_N = \theta \underline{s}_N + \underline{n}_N^{(1)} \tag{2a}$$

and

$$\underline{y}_N = \underline{n}_N^{(2)} \tag{2b}$$

where $\underline{n}_N^{(i)}$, i = 1,2, are mutually independent i.i.d. noise sequences possessing common univariate p.d.f. $f(\cdot)$ as described above. Finally, in the case of a dual-channel detector with a common input signal we observe the two sequences,

$$\underline{x}_N = \theta \underline{s}_N + \underline{n}_N^{(1)} \tag{3a}$$

and

$$\underline{y}_N = \theta \underline{s}_N + \underline{n}_N^{(2)} \tag{3b}$$

with $\underline{n}_N^{(i)}$, i = 1,2, again mutually independent i.i.d noise sequences possessing common univariate p.d.f. $f(\cdot)$ continuous and symmetric about the origin.

Each of the above detection situations can then be formulated as the hypothesis test

$$\begin{aligned} H_0 &: \theta = 0 && \text{(noise only)} \\ \text{vs.} \quad H_1 &: \theta > 0 && \text{(signal in addition to noise present)} \end{aligned} \tag{4}$$

The detection criteria assumed is that of Neyman-Pearson according to which the detection probability or power $\beta_N(\theta)$ is to be maximized subject to the constraint that the false alarm probability not exceed a fixed level α_N. From the Neyman-Pearson lemma [1], [2], the tests of interest then take the form of a threshold test on an appropriate statistic defined for each sample size N. The statistic T_N is then compared to a decision threshold Λ_0. If $T_N < \Lambda_0$ we announce the null hypothesis H_0 as acting while if $T_N \geq \Lambda_0$ we announce the alternative. The false alarm probability is then given by

$$\alpha_N = P_{H_0}\{T_N \geq \Lambda_0\} \tag{5}$$

while the power or detection probability β_N is determined as the probability of this same event when H_1 is known to be acting, i.e.,

$$\beta_N(\theta) = P_{H_1}\{T_N \geq \Lambda_0\} \; . \tag{6}$$

A useful role in what follows is played by the locally most powerful (LMP) detector. In particular, let $\beta_N(\theta)$ denote the power function of an arbitrary detector with fixed false alarm probability α_N where N is, of course, the sample size. Then the detector structure with power function $\beta_N^*(\theta)$ is defined as the LMP detector if

$$\left.\frac{\partial \beta_N^*(\theta)}{\partial \theta}\right|_{\theta=0} \geq \left.\frac{\partial \beta_N(\theta)}{\partial \theta}\right|_{\theta=0} \tag{7}$$

uniformly in N. Under appropriate regularity conditions on the p.d.f. f(•), which we assume to be satisfied, this is the equivalent to maximizing the detection probability in some sufficiently small neighborhood of $\theta = 0$. The LMP detector can then be found by an application of the extended form of the Neyman-Pearson lemma (cf.[1], pp. 83-88). More specifically, consider the single-channel detection situation based upon the sequence of observables $\{\underline{v}_N\}$. The LMP test is then obtained for each N as the threshold test

$$\left.\frac{\partial L(\underline{v}_N\, \theta)}{\partial \theta}\right|_{\theta=0} \mathop{\gtrless}_{H_0}^{H_1} \Lambda_0 \tag{8}$$

where $L(\underline{v}_N|\theta)$ is the likelihood ratio conditioned on a fixed value θ under the alternative, i.e.,

$$L(\underline{v}_N|\theta) \triangleq \frac{p_{H_1}(\underline{v}_N|\theta)}{p_{H_0}(\underline{v}_N)} \tag{9}$$

and Λ_0 in (8) is a generic decision threshold chosen to achieve the desired false alarm probability α_N. Similar comments apply in the dual channel case where the observables are represented by the bivariate sequence $\{\underline{x}_N, \underline{y}_N\}$.

As noted previously, we make exclusive use of A.R.E. as a performance criteria. The A.R.E. is particularly appropriate in the case of weak signal detection when large sample sizes are available. Basically it provides an indication of the relative sample sizes required by two detectors to maintain identical false alarm and detection probabilities in the presence of vanishingly small input SNR.

There exists a simple expression for A.R.E. originally due to Pitman [3] and later extended and generalized by Noether [4] with an exposition given in the book by Fraser [5]. In particular under certain regularity conditions the A.R.E. of one detector with respect to another is given as the ratio of their efficacies, where the efficacy of a detector D_i based upon the sequence of decision statistics $\{T_N^{(i)}\}$ is given by

$$E_{D_i} = \lim_{N\to\infty} \frac{1}{N} \left[\frac{\left.\frac{\partial^m}{\partial\theta^m} E_\theta\{T_N^{(i)}\}\right|_{\theta=0}}{\sqrt{\mathrm{Var}_0\{T_N^{(i)}\}}} \right]^{1/m\nu} \tag{10}$$

where $\mathrm{Var}_0\{T_N^{(i)}\}$ is the variance of the test statistic under H_0, m is the order of the first nonzero derivative of $E_\theta\{T_N^{(i)}\}$ evaluated at $\theta = 0$ and ν is a constant. In most engineering applications it will be the case that $m\nu = 1/2$. The A.R.E. of Detector_1 with respect to another detector D_2 is then given by

$$\text{A.R.E.}_{1,2} = \frac{E_{D_1}}{E_{D_2}} .$$

In what follows it will be convenient to have available a number of parameterized non-Gaussian p.d.f.'s for the additive background noise. These will prove particularly useful in the evaluation of the relative performance of the various adaptive detector structures to be described. We will make use of three classes of parameterized non-Gaussian p.d.f.'s. These three classes of unimodal and symmetric p.d.f.'s are generalizations of the Gaussian, Cauchy, and beta distributions respectively as described by Miller and Thomas [6].

A. <u>Generalized Gaussian Noise</u>

This class of parameterized p.d.f.'s are described according to

$$f(x)=[\alpha\eta(\alpha,\beta)/2\Gamma(1/\alpha)]\exp\{-[\eta(\alpha,\beta)|x|]^\alpha\} \tag{11}$$

where we have defined

$$\eta(\alpha,\beta) \triangleq \beta^{-1}[\Gamma(3/\alpha)/\Gamma(1/\alpha)]^{\frac{1}{2}}$$

with $\alpha>0$ describing the exponential rate of decay, β is a positive quantity representing a scale parameter and $\Gamma(\cdot)$ is the gamma function. The variance of the associated noise distribution is easily seen to be given by $\sigma^2=\beta^2$. For $\alpha=2$ we have the Gaussian distribution

$$f(x) = \frac{1}{\sqrt{2\pi}\sigma} \exp\{ -\frac{1}{2} \frac{x^2}{\sigma^2} \} \tag{12}$$

while for $\alpha = 1$ we have the Laplace distribution

$$f(x) = \frac{1}{2\sigma} \exp\{ -\sqrt{2} \, |\frac{x}{\sigma}| \} . \tag{13}$$

According to Algazi and Lerner [7] values of α in the range $0.1<\alpha<0.6$ have been found representative of certain atmospheric impulsive noise phenomenon. Numerical evaluation of the corresponding likelihood ratio and LMP detector structures for the single channel case are provided in [6], [8]. It is well known, for example, that the LMP detector structure for the case $\alpha=1$ is the sign detector or hard-limiting correlator described in terms of the threshold test

$$T_N(\underline{v}_N) = \sum_{i=1}^{N} s_i \, \mathrm{sgn}(v_i) \underset{H_0}{\overset{H_1}{\gtrless}} \Lambda_0 \tag{14}$$

where

$$\mathrm{sgn}(s) = \begin{cases} +1 \; ; \; x > 0 \\ 0 \; ; \; x = 0 \\ -1 \; ; \; x < 0 \end{cases} \tag{15}$$

Furthermore, as shown in the preceding chapter, the A.R.E. of the sign detector with respect to the linear correlator or matched filter detector is

$$\text{A.R.E.}_{sd,\ell d} = 4\sigma^2 f^2(0) \tag{16}$$

B. Generalized Cauchy Noise

This class of parameterized noise distributions is described in terms of the p.d.f.

$$f(x)=[\alpha\eta(\alpha,\beta)/2\Gamma(1/\alpha)][\nu^{-\frac{1}{2}}\Gamma(\nu+1/\alpha)/\Gamma(\nu)]\{1+[\eta(\alpha,\beta)|x|]^{\alpha}/\nu\}^{-(\nu+1/\alpha)} \quad (17)$$

where again β represents a scale parameter and α,ν are positive parameters completely defining this distribution. The generalized Cauchy distribution exhibits algebraic decay of the tails vis-à-vis the exponential decay associated with the generalized Gaussian distribution. For $\alpha=2$ and $\nu=1/2$ we have the Cauchy distribution

$$f(x) = \frac{1}{\pi\beta}\frac{1}{1+(x/\beta)^2} \quad (18)$$

It is easily shown that in the limit as $\nu\to\infty$ the distribution in (17) approaches the generalized Gaussian distribution of (11) with characteristic exponent α. The variance associated with the generalized Cauchy distribution is given by

$$\sigma^2 = \beta^2[\nu^{2/\alpha}\Gamma(\nu-2/\alpha)|\Gamma(\nu)] \quad (19)$$

which exists only for $\alpha\ \nu > 2$. The generalized Cauchy distribution can be shown to provide a useful model for a variety of non-Gaussian impulsive noise phenomenon.

C. Generalized Beta Noise

Unlike the generalized Gaussian and Cauchy distributions defined for all real x, the generalized beta distribution is confined to a finite interval. In particular, the p.d.f. in this case is given by

$$f(x)=[\alpha\eta(\alpha,\beta)/2\Gamma(1/\alpha)][\nu^{-\frac{1}{2}}\Gamma(\nu+1+2/\alpha)/\Gamma(\nu+1+1/\alpha)].\ \{1-[\eta(\alpha,\beta)|x|]^{\alpha}/\nu\}^{\nu+1/\alpha} \quad (20)$$

for $|x|\le\nu^{1/\alpha}/\eta(\alpha,\beta)$ where α,ν are positive parameters which completely define this distribution. Again it can be shown that as $\nu\to\infty$ the distribution of (20) approaches the generalized Gaussian distribution of (11) with characteristic exponent α. The variance associated with the generalized beta distribution is given by

$$\sigma^2=\beta^2[\nu^{2/\alpha}\Gamma(\nu+1+2)/\Gamma(\nu+1+4/\alpha)] \tag{21}$$

which is finite for all positive α and ν.

III. ADAPTIVE COINCIDENCE PROCEDURES

The adaptive detector structures considered in this section are direct outgrowths of conventional optimum linear and/or quadratic filtering theory suggested by the robust performance of coincidence procedures in general. As stated previously, a coincidence procedure is one that depends only upon the polarities of the observations. Coincidence procedures such as the sign detector defined by (14) or the polarity coincidence correlator (PCC) have previously been considered [9]-[11], for the detection of weak signals in noise of uncertain origin when the noise samples are assumed i.i.d. and the sample size or observation interval is large. These detectors can be easily implemented using binary devices and provide reasonably efficient performance (at least asymptotically) in non-Gaussian broadtailed noise environments. In the presence of Gaussian noise, on the other hand, they are relatively inefficient compared to optimum linear detector structures.

Many detection situations are characterized by time-varying or otherwise unknown noise statistics in which deviations from assumed nominal Gaussian noise conditions exist only a fraction of the time. In such situations it would be desirable to adapt the receiver structure to the prevailing noise conditions. For example, under nominal Gaussian noise conditions the receiver should be linear while under extreme non-Gaussian deviations a nonlinear structure, such as a coincidence detector, would be more appropriate. The adaptive procedures described here are based on this simple observation. In particular, for each of the detection situations under consideration the receiver implements a test statistic which is a suitably weighted linear combination of an optimum linear test statistic and an optimum coincidence test statistic. By appropriate adjustment of the weights, the resulting detector structures attempt to combine the best features of linear matched filtering and coincidence correlation resulting in an asymptotic performance which is an improvement over either of these detectors alone. The appropriate weights to achieve this performance must then be adaptively adjusted on the basis of past observations.

We will postpone until a later section any discussion of adaptive algorithms for adjusting and updating the weights. Here we concentrate upon the structure and performance of these adaptive coincidence detectors. We begin with a discussion of single-channel detectors.

A. Single-Channel Detectors

The adaptive detector structure in this case is illustrated in Fig. 1 where the test statistic is shown to be given by

$$T_N(\underline{v}_N) = \gamma\langle\underline{v}_N, \underline{a}_N\rangle + (1-\gamma)\langle \mathrm{SGN}\ \underline{v}_N, \underline{a}_N\rangle \quad . \qquad (22)$$

Here $\langle \cdot\ \ \cdot \rangle$ represents the ordinary Euclidean inner product, $\mathrm{SGN}\ \underline{v}_N = (\mathrm{sgn}\ v_1, \mathrm{sgn}\ v_2, \ldots, \mathrm{sgn}\ v_N)^T$, $\underline{a}_N$ is a filtering sequence† and γ is a real quantity to be specified. Thus, the final test statistic or decision variable is a linear combination of a linear filter and a coincidence detector. By optimum choice of γ we hope to realize the advantages of either structure while suffering the disadvantages of neither. It is to be noted that the resulting detector no longer provides distribution-

† Note that the choice $\underline{a}_N = \lambda \underline{s}_N$ with λ a scalar represents the matched filter.

free false alarm probability. Nevertheless it should prove useful in a number of communications applications.

Under previously stated symmetry assumptions on the noise distribution $f(\cdot)$ it follows that $E_0\{T_N(\underline{v}_N)\}=0$, while

$$E_\theta\{T_N(\underline{v}_N)\}=\gamma\theta<\underline{s}_N,\underline{a}_N>+(1-\gamma)<E_\theta\{SGN\ \underline{v}_N\},\underline{a}_N> \tag{23}$$

where $E_\theta\{SGN\ \underline{v}_N\}$ is the N-vector with components $E_\theta\{sgn(\theta s_i+n_i)\}$, $i=1,2,\ldots,N$.

Now

$$E\{sgn(\theta s_i+n_i)\} = Pr\{n_i>-\theta s_i\}-Pr\{n_i<-\theta s_i\}= 1-2F(-\theta s_i) \tag{24}$$

with $F(\cdot)$ the cumulative distribution function (c.d.f.) of the noise. Expanding $F(-\theta s_i)$ in a Taylor series expansion about the origin we have

$$F(-\theta s_i) = F(0)-\theta s_i f(0)+0(\theta^2) \tag{25}$$

with $f(x) = dF(x)/dx$. Making use of the symmetry assumptions on $f(\cdot)$ it follows that

$$E_\theta\{sgn(\theta s_i+n_i)\}=2\theta s_i f(0)+0(\theta^2) \tag{26}$$

and hence

$$\left.\frac{\partial E_\theta\{T_N(\underline{v}_N)\}}{\partial\theta}\right|_{\theta=0} = [\gamma+2(1-\gamma)f(0)]<\underline{s}_N,\underline{a}_N> . \tag{27}$$

Likewise it is easily established that

$$Var_0\{T_N(\underline{v}_N)\} = [\sigma^2\gamma^2+2\gamma(1-\gamma)p_o+(1-\gamma)^2]<\underline{a}_N , \underline{a}_N> \tag{28}$$

where σ^2 is the noise variance and p_o is a quantity defined according to

$$p_o \triangleq \int_{-\infty}^{\infty} |x| f(x)dx \tag{29}$$

The corresponding efficacy for this adaptive detector structure is then

$$E_{ad}=\lim_{N\to\infty} \frac{[\gamma+2(1-\gamma)f(0)]^2}{\gamma^2\sigma^2+2\gamma(1-\gamma)p_o+(1-\gamma)^2}\ \frac{<\underline{s}_N,\underline{a}_N>^2}{N<\underline{a}_N,\underline{a}_N>} \tag{30}$$

which reduces to well-known results in the special cases $\gamma=0$ or 1. In the general case, we seek to choose γ to maximize the efficacy E_{ad}. The optimization problem is facilitated by the fact that, as (30) indicates, the optimum choice of γ is independent of the filtering sequence $\underline{a}_N$. While the optimum γ is easily found by straighforward calculus, a detailed derivation of the optimum γ value is provided in the Appendix. This derivation will be useful later in discussing adaptive algorithms for adjusting γ. At any rate, the optimum choice of γ is shown to be given by

$$\gamma_{opt} = \frac{1}{1+a\sigma} \tag{31}$$

where

$$a \triangleq \frac{2\sigma f(0) - p_o/\sigma}{1-2(p_o/\sigma)\sigma f(0)} \tag{32}$$

It is instructive to observe at this point that according to the Schwartz inequality

$$p_o = \int_{-\infty}^{\infty} |x| f(x)dx \leq \{\int_{-\infty}^{\infty} x^2 f(x)dx\}^{\frac{1}{2}} \{\int_{-\infty}^{\infty} f(x)dx\}^{\frac{1}{2}} = \sigma \tag{33}$$

and hence, $0 \leq (p_o/\sigma) \leq 1$. It follows, in particular, that $a \geq 0$ for

$$(p_o/\sigma)^2 \leq 4\sigma^2 f^2(0) \leq (p_o/\sigma)^{-2} \tag{34}$$

and thus $0 \leq \gamma_{opt} \leq 1$ in this range.

Furthermore, assuming the matched filter is employed, i.e., $\underline{a}_N = \lambda \underline{s}_N$, the optimum choice of γ results in

$$E_{ad} = \lim_{N\to\infty} \left[\frac{1-4(p_o/\sigma)\sigma f(0) + 4\sigma^2 f^2(0)}{1-(p_o/\sigma)^2} \right] \frac{\langle \underline{s}_N, \underline{s}_N \rangle}{N\sigma^2} . \tag{35}$$

The corresponding efficacy for the linear matched filter, on the other hand, is given by

$$E_{\ell d} = \lim_{N\to\infty} \frac{\langle \underline{s}_N, \underline{s}_N \rangle}{N\sigma^2} . \tag{36}$$

The A.R.E. of the adaptive detector vis-á-vis the linear matched filter is then clearly

$$\text{A.R.E.}_{ad,\ell d} = \frac{1-4(p_o/\sigma)\sigma f(0)+4\sigma^2 f^2(0)}{1-(p_o/\sigma)^2} \quad . \tag{37}$$

A sketch of this quantity is illustrated in Fig. 2 as a function of $4\sigma^2 f^2(0)$ and parameterized by the quantity $(p_o/\sigma)^2$. Also included is the A.R.E. with respect to the sign test or hard-limiting correlator which is represented as a straight line of unit slope. Observe that the performance of the adaptive detector is always superior to that of either the linear matched filter or the hard-limiting correlator alone. The actual performance improvement depends only upon the two critical parameters $4\sigma^2 f^2(0)$ and $(p_o/\sigma)^2$. For example, if $4\sigma^2 f^2(0)=1$, the expected improvement over either† the linear or hard-limiting correlator is

$$\Delta = \frac{2}{1+(p_o/\sigma)} \tag{38}$$

which approaches 3dB for small (p_o/σ) corresponding to noise with light-tailed distribution.

The values of $4\sigma^2 f^2(0)$ and $(p_o/\sigma)^2$ for several selected noise distributions are given in Table 1. Also included in this table is the quantity $8\sigma^2 \{\int_{-\infty}^{\infty} f^2(x)dx\}^2$ which will play a fundamental role in later discussions of dual-channel detectors with a reference noise channel. Observe that for the Gaussian distribution the quantities $4\sigma^2 f^2(0)$ and (p_o^2/σ^2) are identical while for the Laplace distribution they are reciprocal. It is easily shown that $\gamma_{opt}=1$ for the Gaussian distribution while $\gamma_{opt}=0$ for the Laplace distribution. This agrees with well-known results in these two special cases. That is, the linear receiver is locally (and globally) optimum for Gaussian noise while the hard-limiter provides a locally optimum receiver for Laplace noise.

†Note from Fig. 2 that the performance of the linear and hard-limiting receivers is identical for $4^2\sigma^2 f^2(0)=1$.

The quantities $4\sigma^2 f^2(0)$ and (p_o^2/σ^2) are plotted in Figure 3 for the generalized Gaussian distribution as a function of the characteristic exponent α. This distribution, of course, coincides with the Gaussian and Laplace distributions for the choices $\alpha=2$ and $\alpha=1$ respectively. The improvement in A.R.E. over either the linear or hard-limiting correlator can be substantial, particularly for small values of α. Similar plots of these quantities for the generalized Cauchy and generalized beta distributions can be found in Figs. 4 and 5 respectively.

It is instructive to compare the performance obtained with this nonlinear receiver to that of the locally optimum or locally most powerful detector defined in the preceding section. In particular, the LMP detector as illustrated in Fig. 6 implements the threshold test.

$$T_N(\underline{v}_N) = \sum_{i=1}^{N} a_i \; g_{\ell o}(v_i) \underset{H_0}{\overset{H_1}{\gtrless}} \Lambda_0 \tag{39}$$

where the zero-memory nonlinear characteristic $g_{\ell o}(\cdot)$ is given by

$$g_{\ell o}(x) = -\frac{d \ln f(x)}{dx} \tag{40}$$

and again $f(\cdot)$ is the p.d.f. of the additive noise. It can be shown (cf. [6],[8] that the efficacy of the LMP test under the assumption $\underline{a}_N=\lambda\underline{s}_N$ is given by

$$E_{\ell o} = \lim_{N\to\infty} \sigma^2 \left[\int_{-\infty}^{\infty} \frac{[f'(x)]^2}{f(x)} dx \right] \frac{\langle \underline{s}_N, \underline{s}_N \rangle}{N} \tag{41}$$

The A.R.E. of the adaptive nonlinear detector with respect to the LMP test is then

$$\text{A.R.E.}_{ad,\ell o} = \rho^{-1} \left[\frac{1-4(p_o/\sigma)\sigma f(0)+4\sigma^2 f^2(0)}{1-(p_o/\sigma)^2} \right] . \tag{42}$$

where

$$\rho \triangleq \sigma^2 \int_{-\infty}^{\infty} \frac{[f'(x)]^2}{f(x)} \, dx \quad . \tag{43}$$

For example, with generalized Gaussian noise we have

$$\rho = \frac{\alpha^2 \Gamma(3/\alpha)\Gamma(2-1/\alpha)}{\Gamma^2(1/\alpha)} \tag{44}$$

provided $\alpha > 1/2$. The correspondence A.R.E.$_{ad,\ell o}$ is plotted in Fig. 7 as a function of the characterisitc exponent α. Observe that A.R.E.$_{ad,\ell o}$=1 for α=1.0 and α=2.0 while remaining close to unity for a wide range of α values. For generalized Cauchy noise we find

$$\rho = (\alpha\nu+1)^2 \frac{\Gamma(3/\alpha)\Gamma(2-1/\alpha)\Gamma(\nu-2/\alpha)\Gamma(\nu+1/\alpha)\Gamma(\nu+2/\alpha)}{\Gamma^2(1/\alpha)\Gamma^2(\nu)\Gamma(\nu+2+1/\alpha)} \tag{45}$$

subject to the conditions $\alpha > 1/2$ and $\alpha\nu > 2$. The corresponding A.R.E. is plotted in Fig. 8 as a function of the characteristic parameter α for various choices of ν. Observe again that the A.R.E. is close to unity over a wide range of α and ν values. Finally, in the case of generalized beta noise we have

$$\rho = (\alpha\nu+1)^2 \frac{\Gamma(3/\alpha)\Gamma(2-1/\alpha)\Gamma(\nu-1+1/\alpha)\Gamma^2(\nu+1+2/\alpha)}{\Gamma^2(1/\alpha)\Gamma(\nu+1)\Gamma(\nu+1+1/\alpha)\Gamma(\nu+1+4/\alpha)} \tag{46}$$

where now $\alpha > 1/2$ and $\nu+1/\alpha>1$. Corresponding values of A.R.E.$_{ad,\ell o}$ are plotted in Fig. 9 as a function of the characteristic parameter α for various choices of ν. We see again that the A.R.E. is close to unity for a wide range of α values, at least provided $\nu>1$. From small ν values the p.d.f. is non-zero only within a relatively small interval on the real line. Neither the linear matched filter nor the hard-limiting correlator are particularly effective in this case. As ν increases, on the other hand, the support of $f(\cdot)$ increases resulting in broader tails and the adaptive detector becomes relatively efficient vis-à-vis the LMP detector structure.

In summary then, it appears that the proposed nonlinear detector structure will provide improved performance over either the linear or hard-limiting correlator alone. Furthermore, this improvement is completely characterized in terms of the two parameters $4\sigma^2 f^2(0)$ and $(p_o/\sigma)^2$. To achieve this performance, however, an adaptive procedure is required to adjust the value of γ depending upon the prevailing noise environment. This question will be taken up in a later section.

B. Dual Channel Detectors with Reference Noise Channel

This case will be considered only briefly as it is not significantly different from the preceding rather detailed development of the single-channel adaptive detector. We draw heavily upon this development in what follows. In particular, we assume that the vector $\underline{v}_N$ of observables is obtained as the difference of the X and Y channel samples, i.e, $\underline{v}_N = \underline{x}_N - \underline{y}_N$. As in (22) and the sequel we implement the threshold test

$$T_N(\underline{v}_N) = \gamma\langle \underline{v}_N, \underline{a}_N\rangle + (1-\gamma)\langle \mathrm{SGN}\ \underline{v}_N, \underline{a}_N\rangle \quad . \tag{47}$$

It then follows that $E_0\{T_N(\underline{v}_N)\} = 0$ while corresponding to (27) we have

$$\left.\frac{\partial E_\theta\{T_N(\underline{v}_N)\}}{\partial\theta}\right|_{\theta=0} = [\gamma + 2(1-\gamma)\int_{-\infty}^{\infty} f^2(x)\,dx]\langle \underline{s}_N, \underline{a}_N\rangle \tag{48}$$

and similarly

$$\mathrm{Var}_0\{T_N(\underline{v}_N)\} = [2\gamma^2\sigma^2 + 2\sqrt{2}\gamma(1-\gamma)q_o + (1-\gamma)^2]\langle \underline{a}_N, \underline{a}_N\rangle \tag{49}$$

where σ^2 is the noise variance in each channel and q_o is defined according to

$$q_o \triangleq \frac{1}{\sqrt{2}} \int_{-\infty}^{\infty}\int_{-\infty}^{\infty} |x|\,f(x-y)f(y)\,dx\,dy \tag{50}$$

It is easily shown that $0<(q_o/\sigma)\le 1$. Corresponding to (30) the efficacy of the dual-channel adaptive detector becomes

$$E_{ad} = \lim_{N\to\infty} \frac{[\gamma + 2(1-\gamma)\int_{-\infty}^{\infty} f^2(x)\,dx]^2}{2\gamma^2\sigma^2 + 2\sqrt{2\gamma}(1-\gamma)q_o + (1-\gamma)^2} \; \frac{\langle \underline{s}_N, \underline{a}_N\rangle^2}{N\langle \underline{a}_N, \underline{a}_N\rangle} \tag{51}$$

The optimum weight is again given by (31) with σ replaced by $\sqrt{2}\,\sigma$ and now

$$a \triangleq \frac{2\sqrt{2}\,\sigma \int_{-\infty}^{\infty} f^2(x)\, dx - q_o/\sigma}{1-2\sqrt{2}(q_o/\sigma)\sigma \int_{-\infty}^{\infty} f^2(x)\,dx} \tag{52}$$

Clearly $0 < \gamma_{opt} < 1$ for

$$(q_o/\sigma)^2 \le 8\,\sigma^2 \left[\int_{-\infty}^{\infty} f^2(x)\,dx\right]^2 \le (q_o/\sigma)^{-2} . \tag{53}$$

The optimum choice of γ results in an A.R.E. with respect to the optimum linear detector given by

$$\text{A.R.E.}_{ad,\ell o} = \frac{1-4\sqrt{2}(q_o/\sigma)\sigma \int_{-\infty}^{\infty} f^2(x)\,dx + 8\sigma^2[\int_{-\infty}^{\infty} f^2(x)\,dx]^2}{1-(q_o/\sigma)^2} \tag{54}$$

This compares with the corresponding expression (37) in the case of single channel detectors. The performance of the adaptive dual-channel detector is then completely specified in terms of the quantities $8\sigma^2 \{\int_{-\infty}^{\infty} f^2(x)\,dx\}^2$ and $(q_o/\sigma)^2$ corresponding to $4\sigma^2 f^2(0)$ and $(p_o/\sigma)^2$ respectively in the single-channel case. Similarly, expressions for the A.R.E. with respect to the corresponding LMP detector analogous to (42) are readily apparent. The details are omitted here.

C. Dual-Channel Detector with Common Input Signal

The observables in this case are given by (3). A common, although definitely suboptimum, statistic proposed for this situation is the correlator (cf. [14],[15])

$$T_N(\underline{x}_N, \underline{y}_N) = \langle \underline{x}_N, \underline{\underline{A}}_N\, \underline{y}_N \rangle$$

$$= \sum_{i=1}^{N} \sum_{j=1}^{N} x_i\, a_{ij}\, y_j \tag{55}$$

and its coincidence counterpart

$$T_N(\underline{x}_N,\underline{y}_N) = \langle \mathrm{SGN}\ \underline{x}_N, \underline{\underline{A}}\ \mathrm{SGN}\ \underline{y}_N \rangle$$
$$= \sum_{i=1}^{N} \sum_{j=1}^{N} \mathrm{sgn}\ x_i\ a_{ij} \mathrm{sgn}\ y_j \quad . \tag{56}$$

Here $\underline{\underline{A}}_N$ is an arbitrary nonsingular and symmetric smoothing matrix. In the case $\underline{\underline{A}}_N = \underline{\underline{I}}_N$ -the N × N identity matrix - the statistic given by (56) reduces to the so-called polarity coincidence correlator (PCC) whose performance has been extensively studied [9]-[11]. If the noise in both channels is i.i.d. Gaussian, the optimum Neyman-Pearson test is the energy detector.

$$S_N(\underline{x}_N,\underline{y}_N) = \|\underline{x}_N + \underline{y}_N\|^2 = \sum_{i=1}^{N} [x_i + y_i]^2 \underset{H_0}{\overset{H_1}{\gtrless}} \Lambda_0 . \tag{57}$$

As demonstrated by Carlyle and Thomas [12], the A.R.E. of the analog correlator in (55) with $\underline{\underline{A}}_N = \underline{\underline{I}}_N$ compared to the energy detector is 0.5. Nevertheless, the ease of implementation of the correlator together with its definite intuitive appeal often favor its use over the energy detector.

With the above in mind, an obvious choice of an adaptive detector structure would be to implement the statistic

$$T_N(\underline{x}_N,\underline{y}_N) = \gamma \langle \underline{x}_N, \underline{\underline{A}}_N \underline{y}_N \rangle + (1-\gamma) \langle \mathrm{SGN}\ \underline{x}_N, \underline{\underline{A}}_N\ \mathrm{SGN}\ \underline{y}_N \rangle \tag{58}$$

with the optimum choice of γ to be determined. Under the assumed symmetry assumptions on the noise p.d.f. $f(\cdot)$ it is easily established that $E_0\{T_N(\underline{x}_N,\underline{y}_N)\}=0$, while corresponding to (27) for the single-channel case we have

$$\left. \frac{\partial^2 E_\theta\{T_N(\underline{x}_N,\underline{y}_N)\}}{\partial\theta^2} \right|_{\theta=0} = 2[\gamma + 4(1-\gamma) f^2(0)] \langle \underline{s}_N, \underline{\underline{A}}_N\ \underline{s}_N \rangle . \tag{59}$$

Similarly, the variance under H_0 is given by

$$\mathrm{Var}_0\{T_N(\underline{x}_N,\underline{y}_N)\} = [\sigma^4 \gamma^2 + 2\gamma(1-\gamma) p_o^2 + (1-\gamma)^2] \mathrm{tr}\ \underline{\underline{A}}_N \underline{\underline{A}}_N^T \tag{60}$$

where tr(·) represents the trace of a matrix and p_o is given by (20). It proves convenient to introduce the inner product†

$$<\underline{\underline{P}}_N, \underline{\underline{Q}}_N> \triangleq \operatorname{tr} \underline{\underline{P}}_N \underline{\underline{Q}}_N^T \tag{61}$$

defined on the space of real N × N matrices (cf. Balakrishnan [13]). The efficacy of the adaptive detector given by (58) is then

$$E_{ad} \lim_{N\to\infty} \frac{[\gamma+4(1-\gamma)f^2(0)]^2}{\sigma^4\gamma+2\gamma(1-\gamma)p_o^2+(1-\gamma)^2} \frac{2<\underline{\underline{A}}_N, \underline{\underline{B}}_N>^2}{N<\underline{\underline{A}}_N, \underline{\underline{A}}_N>}$$

where $\underline{\underline{B}}_N = \underline{s}_N \underline{s}_N^T$ is an N × N signal matrix. For fixed γ the efficacy is maximized by the choice $\underline{\underline{A}}_N = \lambda \underline{\underline{B}}_N$. The optimum γ is again given by (31) with σ replaced by σ^2 and now

$$a \triangleq \frac{4\sigma^2 f^2(0) - (p_o/\sigma)^2}{1-4(p_o^2/\sigma^2)\sigma^2 f^2(0)} \tag{63}$$

Assuming optimum choice of $\underline{\underline{A}}_N$ and γ we find

$$E_{ad} = \lim_{N\to\infty} \left[\frac{1-8(p_o/\sigma)^2\sigma^2 f^2(0)+16\sigma^4 f^4(0)}{1-(p_o/\sigma)^4}\right] \frac{2<\underline{\underline{B}}_N, \underline{\underline{B}}_N>}{N\sigma} \tag{64}$$

Similarly, for the analog correlation detector we have

$$E_{cd} = \lim_{N\to\infty} \frac{2<\underline{\underline{B}}_N, \underline{\underline{B}}_N>}{N\sigma^4} \tag{65}$$

It follows that the A.R.E. of the adaptive detector with respect to the correlation detector is then

$$\text{A.R.E.}_{ad,cd} = \frac{1-8(p_o/\sigma)^2\sigma^2 f^2(0)+16\sigma^4 f^4(0)}{1-(p_o/\sigma)^4} \tag{66}$$

The numerical details are not significantly different from that of the single-channel case to warrant additional discussion at this time.

---

†Hopefully it should cause no confusion in that the same symbol <·,·> is being used for two distinct inner products. It should be clear from the context which one is intended.

IV. ADAPTIVE RANK PROCEDURES

The adaptive detector structures considered in the present section are suggested by the weak signal form of certain nonparametric rank tests known to possess desirable locally optimum properties. In particular, many statistical hypothesis testing problems exhibit certain symmetries which can be successfully exploited to aid in developing logical criteria for the choice of test statistic or, in communications terminology, detector functional. For definiteness, consider the single-channel case with observations defined according to (1) where it is desired to implement the hypothesis test given by (4). Observe that this test is invariant under the group H of continuous, odd, and monotone increasing operations on the data, i.e., the nonparametric hypothesis test is unchanged although the transformed data now possesses different distributions. It seems reasonable then that the outcome of any statistical test procedure should likewise be invariant under the group operations represented by H. A test whose outcome is invariant under an appropriately defined group of transformations of the data is said to be an <u>invariant test.</u> By restricting attention to invariant tests in the present case we are in essence ignoring the irrelevant scale to which the data is measured and considering only the relative algebraic ordering of the data samples.

Fortunately, tests which are invariant with respect to a group H of transformations can be completely characterized in terms of so-called maximal invariants. In particular, a function T with domain X is said to be a <u>maximal invariant</u> if

i.) $T(.)$ is invariant under H, i.e., $T(hx)=T(x)$ for all $x \in X$ and $h \in H$.

and

ii.) $T(x_1)=T(x_2)$ implies $x_2=hx_1$ for some $h\epsilon H$

It can be shown (Lehman[1], Chap. 6) that if H is a group of transformations possessing a maximal invariant T, then any invariant test must depend upon the data only through T. In particular, for the single-channel case Lehman shows that the absolute ranks of the $\underline{v}_N$ sequence in addition to the corresponding polarities of the observations constitute a maximal invariant with respect to the group H of continuous, odd, and monotone increasing transformations. This suggests that, if the detection procedure is to be insensitive to the coordinate system in which the data is measured, a significant data reduction is possible in that the detector need base its decisions only on the absolute ranks and polarities of the observations.

Similarly, for the dual-channel case when a reference noise channel is available the basic hypothesis test (4) is clearly invariant under the group H of continuous, odd, monotone increasing transformations applied to the X and Y channel data separately. Lehman shows that the appropriate maximal invariant is the relative algebraic ranks of the pooled X and Y channel data. Finally, in the dual-channel case with a common input signal the relative algebraic ranks computed separately in the X and Y channels provide an appropriate maximal invariant with respect to the group of continuous, odd, monotone increasing transformations of the X and Y channel data.

The above observations will then form the basis for our choice of detector functional in each of the cases considered. In particular, the data is in each case first reduced to the appropriate rank statistic. Since we will be chiefly concerned with the threshold case, the structure of the detector is taken as that of the appropriate locally most powerful rank test. The significance of this lies in some results of Capon [16],

[17] who has considered the A.R.E. of locally most powerful rank tests for the dual-channel case when an independent reference channel is available. Capon has demonstrated that these detectors are asymptotically efficient on the basis of A.R.E. compared with the optimum Neyman-Pearson detector. Similar results can be shown to hold for locally most powerful rank tests for the single-channel case as well as the dual-channel case with a common signal. There is good reason then to restrict attention to locally most powerful rank detector structures.

In each of the cases considered, the appropriate locally most powerful rank test is easily characterized as a ranking device followed by a linear filtering operation on a reduced binary sequence obtained from the data ranking in a manner to be made precise in what follows. Unfortunately, design of the optimum linear filter requires parametric knowledge of the underlying noise distributions. Nevertheless, a reasonable approach is to retain the structure of the locally most powerful rank detector with the linear filter parameters iteratively adjusted to optimize a suitable performance index. Appropriate adaptive algorithms are discussed in the next section. For the present we concentrate upon the structure and performance of the various locally most powerful rank tests.

A. Single-Channel Detectors

It will prove convenient in what follows to assume a constant or d.c. signal such that $s_i=1$, $i=1,2,\ldots,N$. The case of a known signal vector $\underline{s}_N$ in (1) is easily handled by the techniques of the present section although the details are a bit more complicated. As mentioned previously, any invariant test in this case must be based upon the relative ranks in absolute value as well as the polarities of the sequence $\underline{v}_N$. Let $t_i^+(\underline{v}_N)$ represent the number of samples less than or equal to v_i in absolute value. The N-vector of absolute ranks will be represented by

$\underline{t}_N^+ = (t_1^+, t_2^+, \ldots, t_N^+)^T$ while the polarities are, of course, represented by SGN $\underline{v}_N = (\text{sgn } v_1, \text{ sgn } v_2, \ldots, \text{ sgn } v_N)^T$. The locally most powerful invariant test is then determined as the threshold test.

$$\frac{\partial L(\underline{t}_N^+, \text{ SGN } \underline{v}_N \mid \theta)}{\partial \theta} \underset{H_0}{\overset{H_1}{\gtrless}} \Lambda_0 \tag{67}$$

It can be shown [15],[18] that under the stated assumptions,

$$P_{H_0}\{\underline{t}_N^+, \text{ SGN } \underline{v}_N\} = \frac{1}{2^N N!} \tag{68}$$

i.e., the absolute ranks and polarities are independent and each is equally likely. Furthermore,

$$P_{H_1}\{\underline{t}_N^+, \text{ SGN } \underline{v}_N \mid \theta\} = P_{H_1}\{\underline{z}_N \mid \theta\} = \frac{1}{2^N N!} E_0 \left\{ \prod_{i=1}^{N} \frac{g_\theta(z_{Ni} W_N^{(i)})}{f(W_N^{(i)})} \right\} \tag{69}$$

where $g_\theta(x) \triangleq f(x-\theta)$, $\underline{z}_N = (z_{N1}, z_{N2}, \ldots z_{NN})^T$ is a <u>signed</u> <u>rank-order</u> <u>vector</u> with components

$$z_{Ni} = \begin{cases} +1 & ; \text{ if the i'th smallest sample in absolute value was positive} \\ -1 & ; \text{ otherwise} \end{cases} \tag{70}$$

and finally $0 \le W_N^{(1)} < W_N^{(2)} < \ldots < W_N^{(N)}$ represent the ordered absolute values of a sample of size N drawn from a distribution with p.d.f. $f(\cdot)$.

It follows that the likelihood ratio depends only upon $\underline{Z}_N$ and is given by

$$L(\underline{t}_N^+, \text{ SGN } \underline{v}_N \mid \theta) = L(\underline{z}_N \mid \theta) = E_0 \left\{ \prod_{i=1}^{N} \frac{g_\theta(z_{Ni} W_N^{(i)})}{f(W_N^{(i)})} \right\} . \tag{71}$$

Furthermore, from (67) the locally most powerful signed-rank test (LMPSRT) is then given by

$$T_N(\underline{v}_N) = \langle \underline{b}_N, \underline{z}_N \rangle = \sum_{i=1}^{N} b_{Ni}\, z_{Ni} \underset{H_0}{\overset{H_1}{\gtrless}} \Lambda_0 \tag{72}$$

where

$$b_{Ni} \triangleq E_0 \left\{ \frac{\frac{\partial}{\partial\theta} g_\theta(W_N^{(i)})\Big|_{\theta=0}}{f(W_N^{(i)})} \right\} = E_0\{g_{\ell o}(W_N^{(i)})\} \; ; \quad i=1,2,\ldots,N \tag{73}$$

and $g_{\ell o}(\cdot)$ has been defined previous in (40).

Since the parametric knowledge required to implement the b_{Ni}, $i=1,2,\ldots,N$ is assumed unavailable, we consider linear <u>signed</u> <u>rank-order</u> statistics of the form

$$T_N(\underline{v}_N) = \sum_{i=1}^{N} a_{Ni}\, z_{Ni} = \langle \underline{a}_N, \underline{z}_N \rangle \tag{74}$$

where $\underline{a}_N = (a_{N1}, a_{N2}, \ldots a_{NN})^T$ is an N-vector of constants which is to be adaptively updated on the basis of past observations.

Under the stated symmetry assumption on $f(\cdot)$ it follows that $E_0\{T_N(\underline{v}_N)\}=0$ while in general

$$E_\theta\{T_N(\underline{v}_N)\} = \langle \underline{a}_N, E_\theta\{\underline{z}_N\} \rangle \tag{75}$$

with

$$E_\theta\{\underline{z}_N\} = \Sigma'\, \underline{z}_N\, P_{H_1}\{\underline{z}_N | \theta\} \tag{76}$$

Here the summation Σ' is over all 2^N signed rank-vectors $\underline{z}_N$ and $P_{H_1}\{\underline{z}_N|\theta\}$ is given by (69). Expanding this latter expression in a Taylor series expansion about $\theta=0$ we find†

$$P_{H_1}\{\underline{z}_N|\theta\} = P_{H_0}\{\underline{z}_N\} + \frac{\theta}{2^N} \langle \underline{b}_N, \underline{z}_N \rangle + \underline{O}(\theta^2) \tag{77}$$

where $\underline{b}_N = (b_{N1}, b_{N2}, \ldots b_{NN})^T$ has components given by (73). It follows

---

†Here $\underline{O}(\theta^2)$ is a quantity such that $\lim_{\theta \to 0} ||\underline{O}(\theta^2)||/\theta^2 < \infty$

that

$$E_\theta\{\underline{z}_N\} = E_0\{\underline{z}_N\} + \frac{\theta}{2^N}\{\Sigma' \underline{z}_N \underline{z}_N^T\}\underline{b}_N + \underline{0}(\theta^2). \quad (78)$$

Furthermore, it is easily established that

$$\frac{1}{2^N}\Sigma' \underline{z}_N \underline{z}_N^T = \underline{\underline{I}}_N$$

with $\underline{\underline{I}}_N$ again the N x N identity matrix. Thus observing $E_0\{\underline{z}_N\} = 0$ we have

$$E_\theta\{\underline{z}_N\} = \theta\underline{b}_N + \underline{0}(\theta^2) \quad (79)$$

and hence

$$\left.\frac{\partial E_\theta\{T_N(\underline{v}_N)\}}{\partial\theta}\right|_{\theta=0} = <\underline{a}_N, \underline{b}_N> \quad . \quad (80)$$

Similarly, it is easily established that

$$\mathrm{Var}_0\{T_N(\underline{v}_N)\} = <\underline{a}_N, \underline{a}_N> \quad (81)$$

and hence the efficacy of the adaptive linear signed rank-order test is

$$E_{ad} = \lim_{N\to\infty} \frac{<\underline{b}_N, \underline{a}_N>^2}{N<\underline{a}_N, \underline{a}_N>} \quad . \quad (82)$$

The locally optimum linear signed rank-order test is then defined by the choice $\underline{a}_N = \lambda\underline{b}_N$ which then coincides with the LMPSRT. This latter structure results in an A.R.E. of unity with respect to the Neyman-Pearson detector. Consequently by choosing $\underline{a}_N$ to maximize an appropriate local measure of SNR an asymptotically efficient structure results. This will be the key to the adaptive implementation discussed in the next section.

B. Dual-Channel Detectors with Reference Noise Channel

This case will be considered only briefly as it is not significantly different from the preceding discussion of single-channel detectors. The details can be found in [15], [18]. In particular, we suppose that the

X and Y channel samples as described by (2) are pooled to form the sequence $\underline{v}_{2N}=(v_1,v_2,\ldots v_{2N})^T$ of dimension 2N. This sequence is then ordered according to algebraic rank. We then consider the linear rank-order statistic of the form

$$T_N(\underline{v}_{2N})=<\underline{a}_M,\underline{z}_M> = \sum_{i=1}^{M} a_{Mi} z_{Mi} \tag{83}$$

where M=2N-1, $\underline{a}_M = (a_{M1},a_{M2},\ldots a_{MM})^T$ is a filtering sequence and $\underline{z}_M=(z_{M1},z_{M2}\ldots,z_{MM})^T$ is the <u>rank-order vector</u> with components of the form

$$z_{Mi} = \begin{cases} 1 \; ; & \text{is the i'th smallest pooled observation} \\ & \text{is an X or signal channel sample} \\ -1 \; ; & \text{otherwise} \end{cases} \tag{84}$$

The reasons for this choice of linear rank-order statistic are described in some detail in [15]. Here it is shown that the efficacy of the adaptive linear rank-order statistic in (83) is given by

$$E_{ad} = \lim_{N\to\infty} \frac{<\underline{a}_M,\underline{\underline{K}}_M\underline{\tilde{b}}_M>^2}{N<\underline{a}_M,\underline{\underline{K}}_M\underline{a}_M>} \tag{85}$$

where $\underline{\underline{K}}_M$ is a symmetrical and nonsingular M x M matrix defined by

$$\underline{\underline{K}}_M = \begin{bmatrix} 1 & & -1/M \\ & \ddots & \\ -1/M & & 1 \end{bmatrix}$$

The M-vector $\underline{\tilde{b}}_M = (\tilde{b}_{M1},\tilde{b}_{M2},\ldots,\tilde{b}_{MM})^T$ has components $\tilde{b}_{Mi} \triangleq b_{2N,i}-b_{2N,2N}$ i = 1,2,...,M where now corresponding to (73)

$$b_{2N,i} = E_0\left\{\frac{\frac{\partial}{\partial\theta} g_\theta(W_{2N}^{(i)})|\theta=0}{f(W_{2N}^{(i)})}\right\} \tag{87}$$

$$= E_0\{g_{\ell o}(W_N^{(i)})\} \quad ; \quad i=1,2,\ldots,2N$$

with $W_{2N}^{(1)} < W_{2N}^{(2)} < \dots < W_{2N}^{(2N)}$ an ordered sample of size 2N drawn from a distribution with p.d.f. f(·).

The problem of choosing $\underline{a}_M$ to maximize the efficacy in (85) is a familiar one, corresponding to the matched filter problem of parametric detection theory. In particular, the maximizing $\underline{a}_M$ is easily shown to be given by

$$\underline{a}_M = \lambda \underline{\underline{K}}_M^{-1} \underline{\underline{K}}_M \underline{\tilde{b}}_M = \lambda \underline{\tilde{b}}_M \tag{88}$$

where λ is an arbitrary nonnegative scalar. Furthermore, it is shown in [15] that this choice coincides precisely with the corresponding locally most powerful rank detector. Thus, we find again, as in the single-channel case, that by maximizing a local measure on SNR a detector structure results which is asymptotically efficient relative to the corresponding Neyman-Pearson detector.

C. Dual-Channel Detector with Common Input Signal

In this case the X and Y channel observations defined according to (3) are independently ordered according to algebraic rank. We then consider linear rank-correlation statistics of the form†

$$T_N(\underline{x}_N, \underline{y}_N) = \langle \underline{\underline{A}}_M, \underline{\underline{Z}}_M \rangle = \sum_{i=1}^{N} a_{M,ij}\, z_{M,ij} \tag{89}$$

where M=N-1, $\underline{\underline{A}}_M$ is an M x M filtering matrix with i,j element $a_{M,ij}$ and $\underline{\underline{Z}}_M$ is an M x M rank-correlation matrix with i,j element defined according to

$$z_{M,ij} = \begin{cases} 1 - 1/N\ ; & \text{if the bivariate sampling results in the i'th smallest X sample paired with the j'th smallest Y sample} \\ -1/N\ ; & \text{otherwise} \end{cases} \tag{90}$$

†Again we are using the inner product $\langle \underline{\underline{A}}_M, \underline{\underline{B}}_M \rangle = \text{tr}\, \underline{\underline{A}}_M \underline{\underline{B}}_M^T$ defined on the space of M x M matrices.

Space precludes complete treatment of this case. Nevertheless, it is shown in [15], [18] that again by maximizing a local measure of SNR a detector structure results which coincides with the corresponding locally most powerful rank detector. The latter detector is in turn asymptotically efficient compared to the corresponding Neyman-Pearson detector.

V. ADAPTIVE ADJUSTMENT PROCEDURES

If the potential performance advantages of the various adaptive detector structures described in the preceding sections are to be realized some procedure must be employed for adapting the corresponding detector parameters to the existing noise environment. While a variety of procedures are available for this purpose, it will not be the intent here to provide an exhaustive comparative analysis of such procedures. Instead we illustrate a fairly simple procedure based upon stochastic gradient techniques. Additional effort is warranted to optimize the choice of adaptive algorithms considering such factors as convergence speed, tracking ability and ease of implementation. We consider first the case of adaptive coincidence procedures.

A. Adaptive Coincidence Procedures

For definiteness we treat the single-channel case with the understanding that similar comments apply to the dual-channel detection situations as well. The adaptive procedure then attempts to iteratively adjust the parameter γ based upon past observations to maximize the SNR of the decision variable $T_N(\underline{v}_N)$ defined by (22). That is, we seek to maximize

$$\mathrm{SNR} \triangleq \frac{E^2\{T_N(\underline{v}_N)\}}{\mathrm{Var}\{T_N(\underline{v}_N)\}} \tag{91}$$

and it is implicitly assumed that a signal is known to be present, i.e., a training period is employed.

It will be convenient in what follows to represent the decision variable as

$$T_N(\underline{v}_N) = \langle \underline{t}_N, \underline{\gamma} \rangle \tag{92}$$

where $\underline{t}_N = (t_{N1}, t_{N2})^T$ with (93)

$$t_{N1} = \frac{1}{\sigma} \langle \underline{v}_N, \underline{a}_N \rangle$$

and

$$t_{N2} = \langle \mathrm{SGN}\, \underline{v}_N, \underline{a}_N \rangle \tag{94}$$

Here the vector $\underline{\gamma}$ has components $(\sigma\gamma, 1-\gamma)$ as defined in the Appendix. The iterative adjustment procedure is facilitated by the observation (cf. Kushner [19]) that the SNR of the decision variable $T_N(\underline{v}_N)$ is maximized by minimizing the mean-square difference between $T_N(\underline{v}_N)$ and an arbitrary positive constant C, i.e., we seek to minimize

$$J(\underline{\gamma}) = E\{[\langle \underline{t}_N, \underline{\gamma} \rangle - C]^2\} \tag{95}$$

by choice of $\underline{\gamma}$. As is well-known (cf. Papoulis [20]), the minimizing $\underline{\gamma}$ is found as solution to the equation

$$E\{[\langle \underline{t}_N, \underline{\gamma} \rangle - C]\, t_N^T\} = 0 \quad . \tag{96}$$

That is, the residual must be uncorrelated with the observable represented by the vector $\underline{t}_N$. The result is, of course,

$$\gamma = C\underline{\underline{K}}^{-1}\underline{m} \tag{97}$$

with $\underline{\underline{K}} = E_\theta\{\underline{t}_N \underline{t}_N^T\}$ and $\underline{m} = E_\theta\{\underline{t}_N\}$. Observe that for small input signals

$$E_\theta\{\underline{t}_N\} \approx \theta \langle \underline{s}_N, \underline{a}_N \rangle \underline{b} \tag{98}$$

where $\underline{b} = (1/\sigma, 2f(0))^T$. This result follows directly from (23) and the sequel. Now assuming † $E_\theta\{\underline{t}_N\underline{t}_N^T\} \approx E_0\{\underline{t}_N\underline{t}_N^T\}$ we have

$$\underline{\underline{K}} = \langle \underline{a}_N, \underline{a}_N \rangle \, \underline{\underline{Q}} \tag{99}$$

where $\underline{\underline{Q}}$ is the 2 x 2 matrix defined in the Appendix. Thus from (97)

$$\underline{\gamma} = \frac{\theta C \langle \underline{s}_N, \underline{a}_N \rangle}{\langle \underline{a}_N, \underline{a}_N \rangle} \underline{\underline{Q}}^{-1} \underline{b} \tag{100}$$

which is the form $\underline{\gamma} = \lambda \underline{\underline{Q}}^{-1} \underline{b}$ and is recognized from the Appendix as the optimum $\underline{\gamma}$.

The iterative adjustment procedure will proceed as follows: let $\underline{\gamma}^0$ be chosen arbitrarily and define

$$\underline{\gamma}^{n+1} = \underline{\gamma}^n - 2^{-1} \alpha_n \underline{g}(\underline{t}_N^n; \underline{\gamma}^n) \quad ; \quad n=0,1,\ldots \tag{101}$$

where $\{\alpha_n\}$ is a sequence of scalars representing step size and $\{\underline{t}_N^n\}$ is a sequence of test statistics obtained on a sequence $\{\underline{v}_N^n\}$ of independent N-blocks of data. Finally, $\underline{g}(\underline{t}_N;\underline{\gamma})$ in (101) represents the gradient with respect to $\underline{\gamma}$ of the argument of the expectation operator on the right-hand side of the defining expression (95) for $J(\underline{\gamma})$. In particular, we have

$$\underline{g}(\underline{t}_N^n;\gamma^n) = \nabla\{[\langle \underline{t}_N^n, \gamma^n \rangle - C]^2\} = 2\{[\langle \underline{t}_N^n, \gamma^n \rangle - C]\underline{t}_N^n\} \quad ; \quad n=1,2 \tag{102}$$

For example, suppose the weights satisfy

$$\sum_{n=0}^{\infty} \alpha_n = \infty \; ; \; \sum_{n=0}^{\infty} \alpha_n^2 < \infty \tag{103}$$

This is merely the Robbins-Monro stochastic approximation procedure [21], [22], which under the stated conditions will converge both in mean-square and with probability one to the unique root of (96). As we have seen, this leads to the optimum $\underline{\gamma}$, i.e., $\gamma = \gamma_{opt}$. Because of the relatively slow

† We are essentially assuming that the signal level is small enough such that higher order terms in θ are neglibible.

convergence of stochastic approximation procedures, however, other choices of the step sequence $\{\alpha_n\}$ may be preferred. This issue is not explored here.

B. Adaptive Rank Procedure

The iterative adjustment procedure will be described only for the single-channel case, where again analogous results hold for the other two cases. Here we seek to maximize the SNR of the decision variable $T_N(\underline{v}_N)$ given by (74). Again we accomplish this by choosing the smoothing sequence $\underline{a}_N$ to minimize the quantity

$$J(\underline{a}_N) = E\{[\langle\underline{a}_N, \underline{z}_N\rangle - C]^2\} \tag{104}$$

with C an arbitrary positive constant. Equivalently, we seek the solution to the regression equation

$$E\{[\langle\underline{a}_N, \underline{z}_N\rangle - C]\underline{z}_N^T\} = \underline{0} \quad . \tag{105}$$

Again the solution is given by

$$\underline{a}_N = C\,\underline{\underline{K}}^{-1}\,\underline{m} \tag{106}$$

with $\underline{\underline{K}} = E_\theta\{\underline{z}_N\underline{z}_N^T\}$ and $\underline{m} = E_\theta\{\underline{z}_N\}$. As in the previous case we assume that $E_\theta\{\underline{z}_N\underline{z}_N^T\} = E_\theta\{\underline{z}_N\underline{z}_N^T\}$ so that

$$\underline{\underline{K}} \approx E_0\{\underline{z}_N\underline{z}_N^T\} = \underline{\underline{I}}_N \tag{107}$$

where we have made use of (78) and the sequel to establish the last equality. Likewise from (79), up to higher order θ terms we have

$$\underline{m} = E_\theta\{\underline{z}_N\} \approx \theta\underline{b}_N \quad . \tag{108}$$

Thus, the unique solution to (105) is given approximately by

$$\underline{a}_N = \theta C \underline{b}_N \tag{109}$$

which is again of the form $\underline{a}_N = \lambda\underline{b}_N$ and is recognized as the form of the LMPSRT.

The iterative adjustment procedure then proceeds as follows: let $\underline{a}_N^0$ be chosen arbitrarily and define

$$\underline{a}_N^{n+1} = \underline{a}_N^n - 2^{-1} \alpha_n \underline{g}(\underline{z}_N^n; \underline{a}_N^n) \quad ; \quad n=0,1,\ldots \tag{110}$$

Here $\{\alpha_n\}$ is again a sequence of scalar step sizes while $\{\underline{z}_N^n\}$ represents a sequence of independent signed-rank vectors computed on the basis of observing the sequence $\{\underline{v}_N^n\}$ of independent N-blocks of data. The appropriate noisy gradient is in this case

$$\underline{g}(\underline{z}_N^n; \underline{a}_N^n) = 2\{[\langle \underline{z}_N^n, \underline{a}_N^n \rangle - C]\underline{z}_N^n\} \quad ; \quad n=1,2,\ldots \tag{111}$$

The previous comments concerning convergence properties and choice of step size continue to apply here. The resulting adaptive detector structure is illustrated in Fig. 10.

VI. SUMMARY

Several adaptive nonparametric detection techniques have been considered. It was shown that there exist fairly simple iterative adjustment procedures for adapting the detector parameters to the unknown prevailing noise environment. These detectors provide relatively efficient performance compared to their parametric counterparts. The adaptive coincidence procedures are particularly simple to implement. Unfortunately, they lose their distribution-free false alarm capabilities. This can be remedied by appropriate adjustment of the decision threshold although this issue was not taken up here. Some results on adaptive threshold adjustment procedures which should have application here are provided in [15].

The adaptive rank detectors are a bit more complicated to implement although potentially offering substantial performance advantage. Here the major implementation difficulty lies in the necessity to order the samples in each successive N-block of data. It is expected that the mixed statistical tests introduced by Feustel and Davisson [23], [24] may offer some advantages in this regard. A <u>mixed statistic</u> is computed by dividing the sample size N into p groups of m samples apiece. A generally rank-like intermediate statistic is computed for each of the p size m subgroups and the final test statistic is computed by summing these p intermediate statistics. The resulting A.R.E.'s compare favorably to the corresponding fully-ranked statistics.

Finally one should mention several other approaches related to the material presented here. Kanefsky and Thomas [25] describe an adaptive scheme which is basically a scheme for adapting the decision threshold in a coincidence detector in an effort to maintain distribution-free false alarm probability. Groginsky, et.al. [26] have considered an adaptive rank procedure for the dual-channel case when an independent noise channel is available. The procedure is similar to the adaptive rank procedures described here. Also Cohen and Kurz [27] have described an adaptive scheme for detection in the presence of noise whose p.d.f. is defined as a mixture distribution. The detector is implemented as a linear convex combination of two ranks statistics. Procedures are given for adapting the weights to the parameters of the mixture distribution. Some simulation results are also described.

APPENDIX. DERIVATION OF OPTIMUM γ VALUE

From Equation (30) it is apparent that for any choice of signal sequence or filter function, the optimum γ is obtained by maximizing the ratio

$$r(\gamma) = \frac{[\gamma+2(1-\gamma)f(0)]^2}{\gamma^2\sigma^2+2\gamma(1-\gamma)p_o+(1-\gamma)^2} \tag{112}$$

Now defining the vectors $\underline{\gamma} = (\sigma\gamma,\ 1-\gamma)^T$ and $\underline{b} = (1/\sigma, 2F(0))^T$, and the 2 x 2 matrix

$$Q = \begin{bmatrix} 1 & p_o/\sigma \\ & \\ p_o/\sigma & 1 \end{bmatrix} \tag{113}$$

This last expression can be written as a ratio of inner products

$$r(\underline{\gamma}) = \frac{\langle \underline{\gamma}, \underline{b} \rangle^2}{\langle \underline{\gamma}, \underline{\underline{Q}}\, \underline{\gamma} \rangle} \tag{114}$$

Due to the positive definiteness of $\underline{\underline{Q}}$ it can always be factored into the product of two nonsingular matrices $\underline{\underline{Q}} = \underline{\underline{P}}^T\, \underline{\underline{P}}$. For example,

$$P = \begin{bmatrix} 1 & 0 \\ & \\ p_o & \sigma\sqrt{1-(p_o/\sigma)^2} \end{bmatrix} \tag{115}$$

suffices. Defining $\underline{\xi} = \underline{\underline{P}}\underline{\gamma}$ we obtain

$$r(\underline{\gamma}) = \frac{\langle \underline{\underline{P}}^{-1}\, \underline{\xi}, \ \underline{b} \rangle^2}{\langle \underline{\xi}, \underline{\xi} \rangle} \tag{116}$$

and by a trivial application of the Schwartz inequality we find the maximum occurs with $\underline{\xi} = \lambda \underline{\underline{P}}^{-T}\underline{b}$ with λ an arbitrary nonzero scalar.

Equivalently, the maximum is achieved with $\underline{\gamma} = \lambda \underline{\underline{Q}}^{-1}\underline{b}$. In this case we have the optimum value

$$r_{opt}(\underline{\gamma}) = \langle \underline{b}, \ \underline{\underline{Q}}^{-1}\underline{b} \rangle = \frac{1}{\sigma^2}\left[\frac{1-4(p_o/\sigma)\sigma f(0)+4\sigma^2 f^2(0)}{1-(p_o/\sigma)^2}\right]. \tag{117}$$

The optimum γ is found as the solution to $\underline{\gamma} = \lambda \underline{\underline{Q}}^{-1}\underline{b}$ with the result

$$\gamma_{opt} = \frac{1}{1+a\sigma} \tag{118}$$

where

$$a \triangleq \frac{2\sigma f(0) - p_o/\sigma}{1-2(p_o/\sigma)\sigma f(0)} \tag{119}$$

Note that a can in fact be negative resulting in γ values greater than unity.

REFERENCES

1. E. L. Lehman, Testing Statistical Hypotheses, Wiley, New York, 1959.

2. D. Middleton, Introduction to Statistical Communication Theory, McGraw-Hill, New York, 1960.

3. E.J.G. Pitman, "Lecture Notes on Nonparametric Statistical Inference", Columbia University, 1948.

4. G. E. Noether, "On a Theory of Pitman", Ann. Math. Stat., Vol. 26, 1955, pp. 64-68.

5. D.A.S. Fraser, Nonparametric Methods in Statistics, Wiley, New York, 1951.

6. J. H. Miller and J. B. Thomas, "Detectors for Discrete-Time Signals in Non-Gaussian Noise", IEEE Trans. Inform. Theory, Vol. IT-18, March 1972, pp. 241-250.

7. V. R. Algazi and R. M. Lerner, "Binary Detection in White Non-Gaussian Noise", M.I.T. Lincoln Lab., Lexington, Mass., Tech. Rept. DS-2138.

8. J. H. Miller, "Detection of Signals in Non-Gaussian Noise", Ph.D. dissertation, Dept. Elect. Eng., Princeton Univ., Princeton, N.J., 1973.

9. M. Kanefsky and J. B. Thomas, "On Polarity Detection Schemes with Non-Gaussian Inputs", Journ. Franklin Inst., Vol. 280, Aug. 1965, pp. 120-138.

10. J. Capon, "Optimum Coincidence Procedures for the Detection of Weak Signals in Noise", IRE Conv. Rec., pt. 4, 1960, pp. 154-166.

11. S. S. Wolf, J. B. Thomas and T. R. Williams, "The Polarity Coincidence Correlator: A Nonparametric Detection Device", IRE Trans. Inform. Theory, vol. IT-18, Jan. 1962, pp. 5-9.

12. J. W. Carlyle and J. B. Thomas, "On Nonparametric Signal Detectors", IEEE Trans. Inform. Theory, vol. IT-10, April 1964, pp. 146-152.

13. A. V. Balakrishnan, "Filtering and Prediction Theory", Chapt. 3 in Communication Theory, ed. by A. V. Balakrishnan, McGraw-Hill, New York, 1968.

14. J. W. Modestino, "Coincidence Procedures for the Detection of Known Signals in Statistically Undefined Background Noise", TR-1, Elect. and Syst. Eng. Dept., Rensselaer Polytechnic Inst., Troy, N. Y., Feb. 1975.

15. J. W. Modestino, "Nonparametric and Adaptive Detection of Dependent Data", Ph.D. dissertation, Dept. of Elect. Eng., Princeton University, Princeton, N.J., 1969.

16. J. Capon, "On the Asymptotic Relative Efficiency of Locally Optimum Detectors", IEEE Trans. Inform. Theory, vol. IT-7, April 1961, pp. 67-72.

17. J. Capon, "Asymptotic Efficiency of Certain Locally Most Powerful Rank Tests". Ann. Math Stat., vol. 32, 1962, pp. 88-100.

18. J. W. Modestino and L. D. Davisson, "Adaptive Detection of Weak Signals in Statistically Undefined Noise Backgrounds", Proc. of Symposium on Computer Processing in Communications, Polytechnic Institute of Brooklyn, New York, April 1969, pp. 629-651.

19. H. Kushner, "Adaptive Techniques for the Optimization of Binary Detection Systems," IEEE, Int. Conv. Rec., vol. 11, pt. 4, 1963, pp. 107-117.

20. A. Papoulis, Probability, Random Variables and Stochastic Processes, McGraw-Hill, New York, 1965, Chap. 11.

21. R. Robbins and S. Munro, "A Stochastic Approximation Method", Ann. Math. Stat., vol. 22, 1951, pp. 400-407.

22. A. Dvoretsky, "On Stochastic Approximation", Proc. 3rd Berkeley Symp. on Math. Stat. and Prob., vol. I,Univ. of Calif. Press, Berkeley, Ca.

23. E. A. Feustel and L, D. Davisson, "The Asymptotic Relative Efficiency of Mixed Statistical Tests", IEEE Trans. Inform. Theory, vol. IT-13, April 1967, pp. 247-255.

24. E. A. Feustel and L. D. Davisson, "On the Efficacy of Mixed Locally Most Powerful One-and-Two-Sample Rank Tests", IEEE Trans. Inform. Theory, vol. IT-14, Sept. 1968, pp. 776-778.

25. M. Kanefsky and J. B. Thomas,"On Adaptive Nonparametric Detection Systems Using Dependent Samples", IEEE Trans. Inform. Theory, vol. IT-11, Oct. 1965, pp. 521-526.

26. H. L. Groginsky, L. R. Wilson and D. Middleton, "Adaptive Detection of Statistical Signals in Noise", IEEE Trans. Inform. Theory, vol. IT-12, July 1966, pp. 337-348.

27. S. Cohen and L. Kurz, "An Efficient Nonparametric Detection Procedure Using a Computer Selection Technique", Proc. of Symposium on Computer Processing in Communications, Polytechnic Institute of Brooklyn, New York, April 1969, pp. 617-628.

DISTRIBUTION	PROBABILITY DENSITY FUNCTION	$4\sigma^2 f^2(0)$	$(p_o/\sigma)^2$	$8\sigma^2\{\int_{-\infty}^{\infty} f^2(x)dx\}^2$
GAUSSIAN	$f(x)=\frac{1}{\sqrt{2\pi\sigma^2}}\exp\{-\frac{1}{2}\frac{x^2}{\sigma^2}\}$	$2/\pi$	$2/\pi$	$2/\pi$
LAPLACE	$f(x)=\frac{1}{\sqrt{2\sigma^2}}\exp\{-\sqrt{2}\,\lvert\frac{x}{\sigma}\rvert\}$	2	1/2	1
GENERALIZED GAUSSIAN	$f(x)=[\alpha\eta(\alpha,\beta)/2\Gamma(1/\alpha)]\exp\{-[\eta(\alpha,\beta)\lvert x\rvert]^{\alpha}\}$	$\frac{\alpha^2\Gamma(3/\alpha)}{\Gamma^3(1/\alpha)}$	$\frac{\Gamma^2(2/\alpha)}{\Gamma(1/\alpha)\Gamma(3/\alpha)}$	$\frac{2^{1-2/\alpha}\alpha^2\Gamma(3/\alpha)}{\Gamma^3(1/\alpha)}$
GENERALIZED CAUCHY	$f(x)=[\alpha\eta(\alpha,\beta)/2\Gamma(1/\alpha)][\nu^{-1/2}\Gamma(\nu+1/\alpha)/\Gamma(\nu)]\{1+[\eta(\alpha,\beta)\lvert x\rvert]^{\alpha}/\nu^{-(\nu+1/\alpha)}$	$\frac{\alpha^2\Gamma(3/\alpha)\Gamma^2(\nu+1/\alpha)\Gamma(\nu-2/\alpha)}{\Gamma^3(1/\alpha)\Gamma^3(\nu)}$	$\frac{\Gamma^2(2/\alpha)\Gamma^2(\nu-1/\alpha)}{\Gamma(1/\alpha)\Gamma(3/\alpha)\Gamma(\nu)\Gamma(\nu-2/\alpha)}$	$\frac{2\alpha^2\Gamma(3/\alpha)\Gamma(\nu-2/\alpha)\Gamma^4(\nu+1/\alpha)\Gamma^2(2\nu+1/\alpha)}{\Gamma^3(1/\alpha)\Gamma^5(\nu)\Gamma^2(2\nu+2/\alpha)}$
GENERALIZED BETA	$f(x)=[\alpha\eta(\alpha,\beta)/2\Gamma(1/\alpha)][\nu^{-1/2}\Gamma(\nu+1+2/\alpha)/\Gamma(\nu+1+1/\alpha)]\{1-[\eta(\alpha,\beta)\lvert x\rvert]^{\alpha}/\nu\}^{\nu+1/\alpha}$	$\frac{\alpha^2\Gamma(3/\alpha)\Gamma^3(\nu+1+2/\alpha)}{\Gamma^3(1/\alpha)\Gamma(\nu+1+4/\alpha)\Gamma^2(\nu+1+1/\alpha)}$	$\frac{\Gamma^2(2/\alpha)\Gamma(\nu+1+2/\alpha)\Gamma(\nu+1+4/\alpha)}{\Gamma(1/\alpha)\Gamma(3/\alpha)\Gamma^2(\nu+1+3/\alpha)}$	$\frac{2\alpha^2\Gamma(3/\alpha)\Gamma^5(\nu+1+2/\alpha)\Gamma^2(2\nu+1+2/\alpha)}{\Gamma^3(1/\alpha)\Gamma^4(\nu+1+1/\alpha)\Gamma(\nu+1+4/\alpha)\Gamma^2(2\nu+1+3/\alpha)}$

Note: $\eta(\alpha\ \beta) \triangleq \beta^{-1}[\Gamma(3/\alpha)/\Gamma(1/\alpha)]^{1/2}$

TABLE 1 Some Typical Noise Probability Distributions and Associated Properties

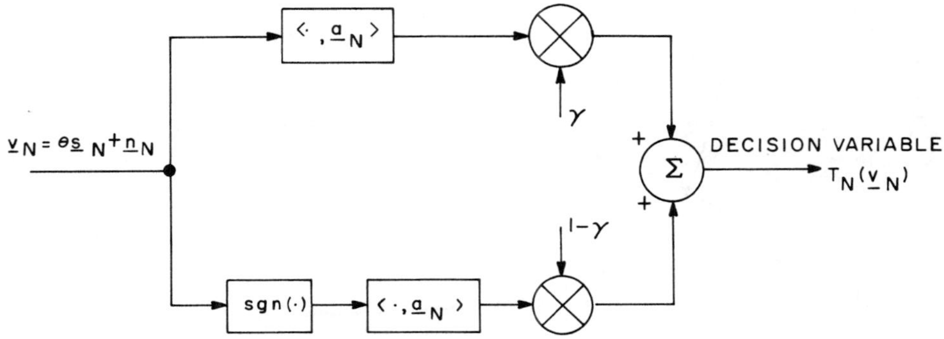

Figure 1 Proposed Nonlinear Adaptive Detector Structure.

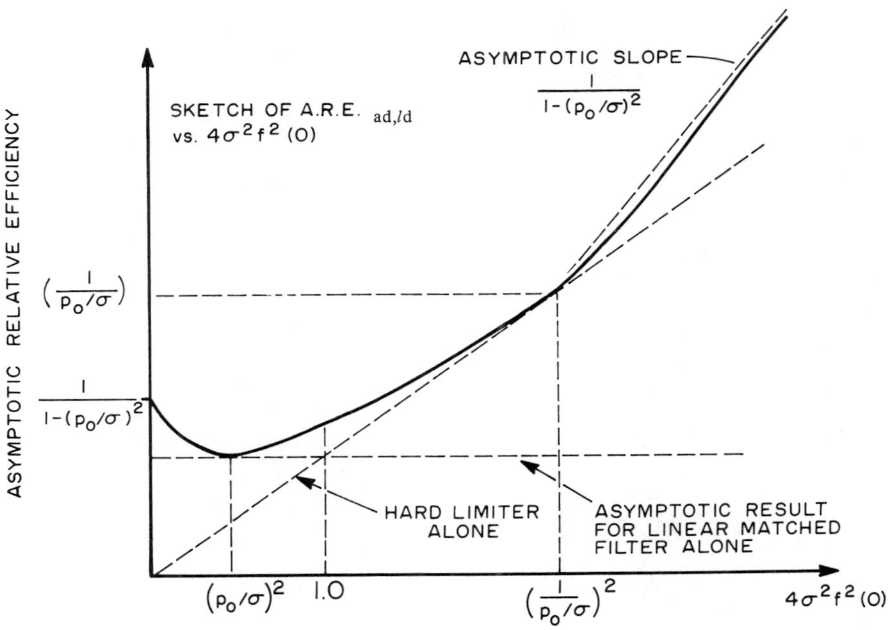

Figure 2 Sketch of A.R.E. of Adaptive Detector with Respect to Linear Detector as a Function of the Quantity $4\sigma^2 f^2(0)$.

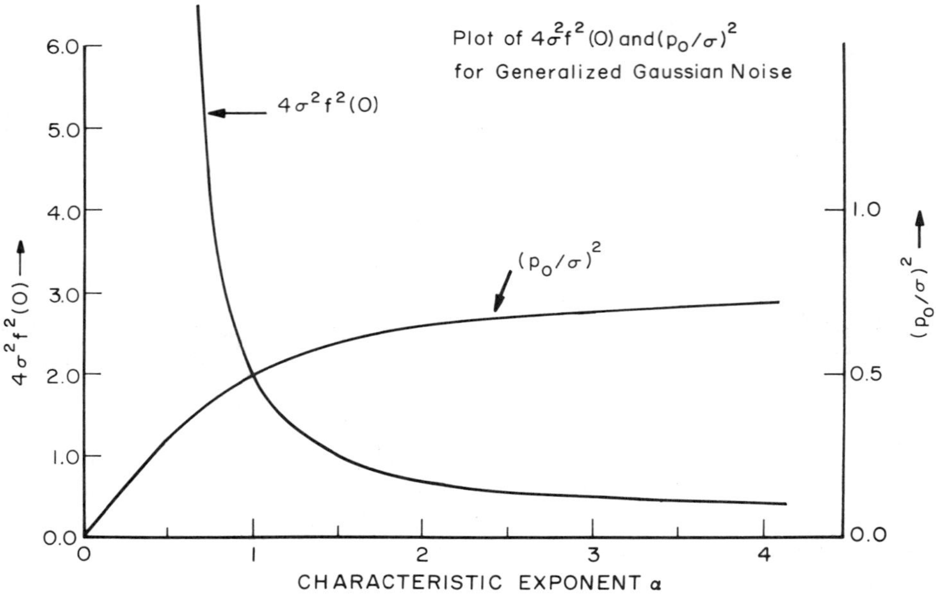

Figure 3 Plot of the Quantities $4\sigma^2 f^2(0)$ and $(p_o/\sigma)^2$ as a Function of the Characteristic Exponent α for Generalized Gaussian Distribution.

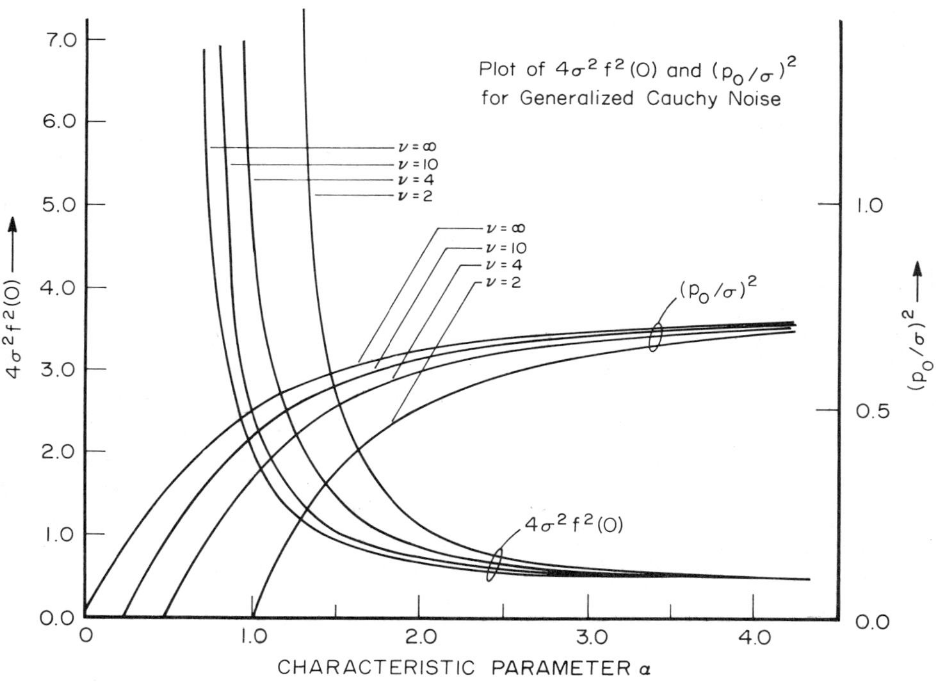

Figure 4 Plot of the Quantities $4\sigma^2 f^2(0)$ and $(p_o/\sigma)^2$ as a Function of the Characteristic Parameter α for Generalized Cauchy Distribution.

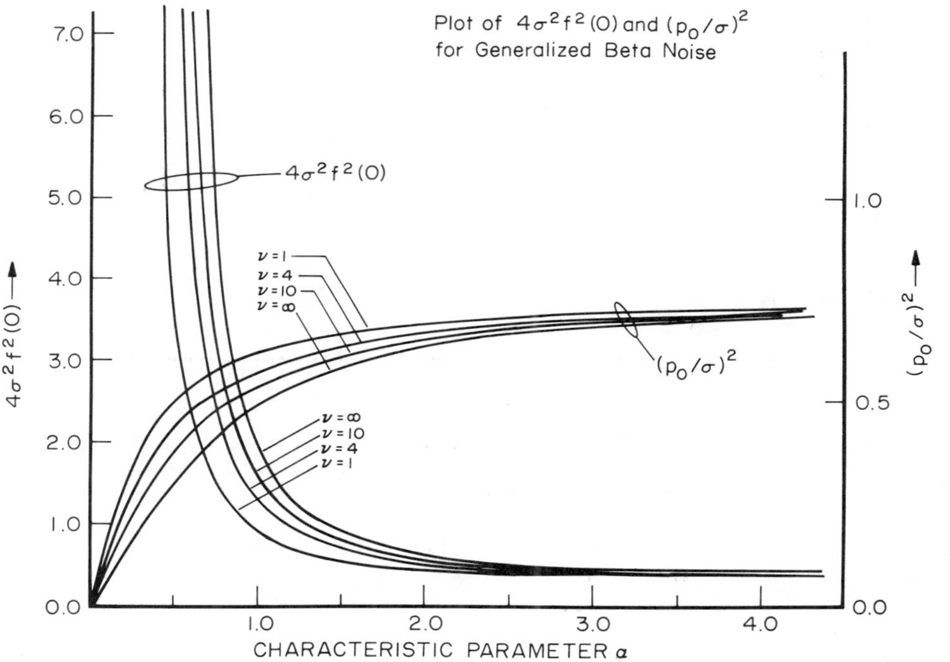

Figure 5 Plot of the Quantities $4\sigma^2 f^2(0)$ and $(p_o/\sigma)^2$ as a Function of the Characteristic Parameter α for Generalized Beta Distribution.

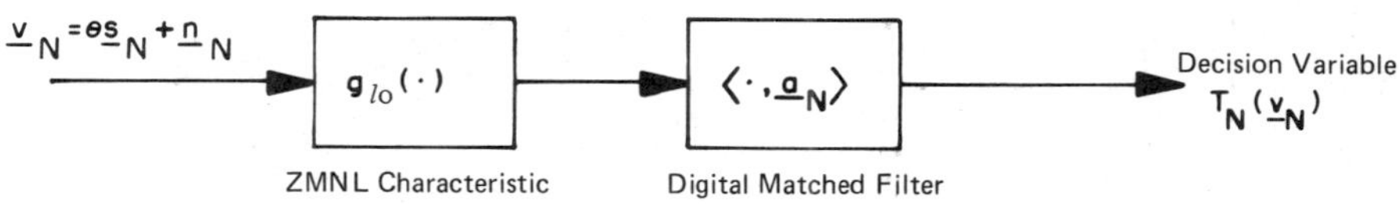

Figure 6 Digital Implementation of Locally Most Powerful Detector Structure.

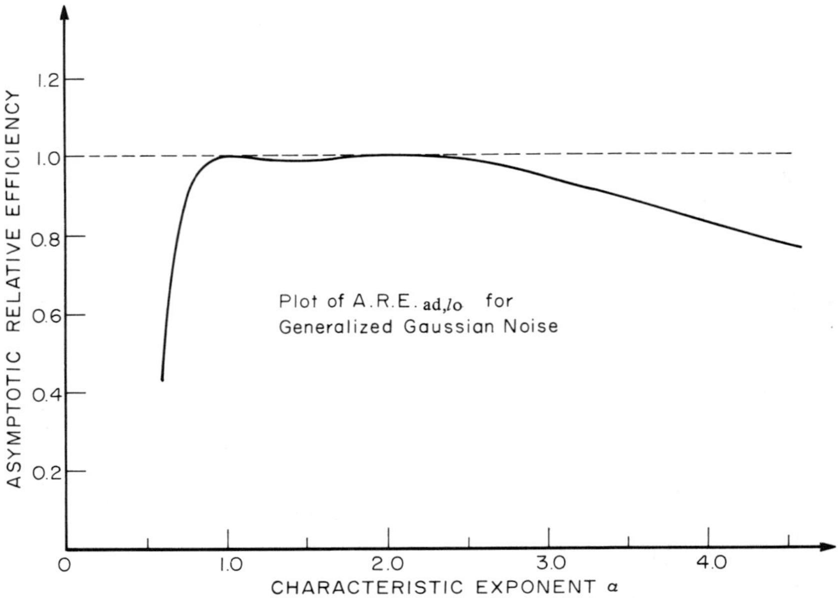

Figure 7 A.R.E. of Adaptive Detector with Respect to the Locally Optimum Detector as a Function of the Characteristic Exponent α for Generalized Gaussian Noise.

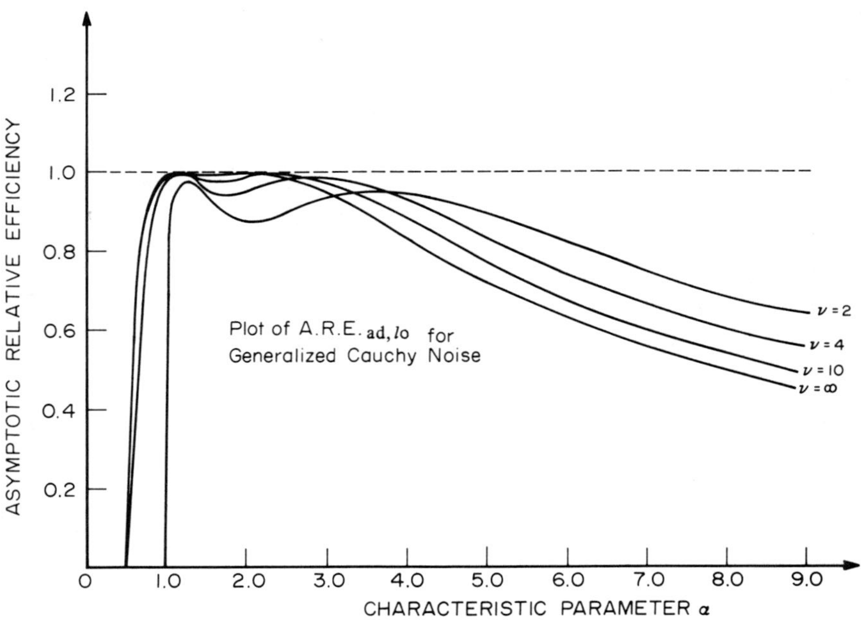

Figure 8 A.R.E. of Adaptive Detector with Respect to the Locally Optimum Detector as a Function of the Characteristic Parameter α for Generalized Cauchy Noise.

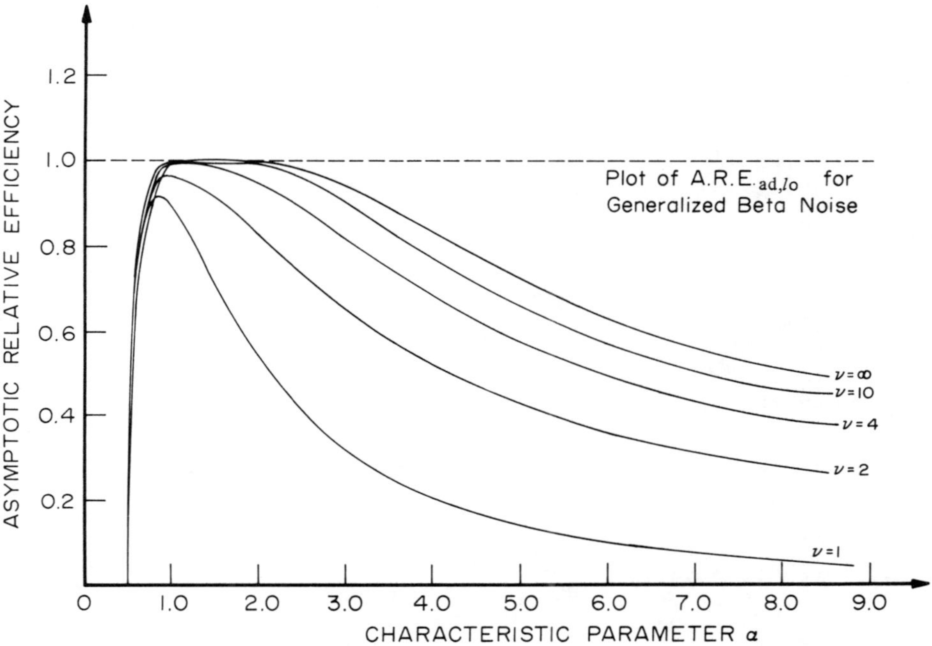

Figure 9 A.R.E. of Adaptive Detector with Respect to the Locally Optimum Detector as a Function of the Characteristic Parameter α for Generalized Beta Noise.

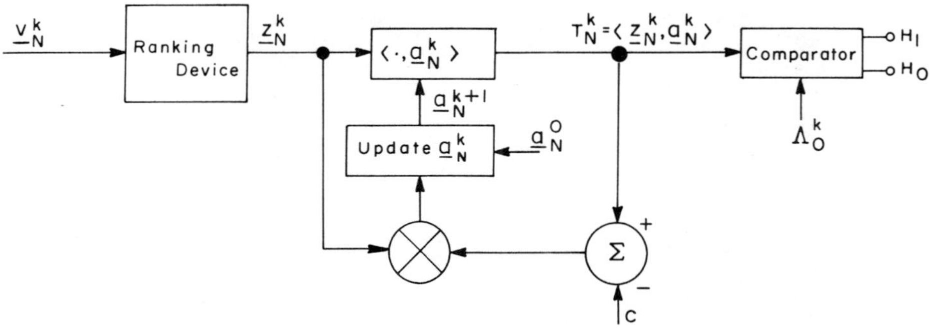

Figure 10 Adaptive Rank Detector for Single-Channel Detection Situation.

Chapter 3

NONPARAMETRIC DETECTORS BASED ON PARTITION TESTS

Ludwik Kurz

Department of Electrical Engineering and Electrophysics
Polytechnic Institute of New York
Brooklyn, New York

I. INTRODUCTION

In this chapter, the problem of detection, when the functional form of the underlying distributions associated with the observables under the different hypotheses is unknown, is considered. The approach taken is to first reduce the observation space by partitioning it on the basis of some general knowledge about a particular class of problems. In particular, a finite number of quantiles is an especially simple and useful set of partitioning the space of observables. The second step involves the selection of the test statistics which are robust (relatively insensitive to the variations of underlying distributions), simple to implement and powerful. In particular, several classes of binary threshold and M-ary "matched filter" type linear and nonlinear m-interval detectors with simple implementation are considered. The M-ary problem is explored and analyzed in terms of efficacies, cross-efficacies and statistic-to-statistic correlation coefficients. It is shown that the m-interval detectors are distribution-free under the hypothesis, asymptotically normal and, in their most powerful form, approach the performance of locally most powerful detectors. Applications of quantile detectors in their linear forms to problems of impulsive and dispersive noise are considered, and they are shown to be powerful, efficient and easy to implement. Two versions of the adaptive form of m-interval detectors are presented: one version is self-adaptive and the other uses a recursive algorithm to estimate the unknown quantiles and the values of p.d.f. at those quantiles. A three-sample version of quadratic quantile detectors is studied using small-signal-analysis techniques. The asymptotic normality of the quadratic tests, based on Whittle's theorem, is proven for shift-of-the-mean and change-of-scale detection models. Based on the asymptotic probability of error, the quadratic three-sample tests are competitive with corresponding linear rank tests in stochastic ordering problems and more efficient in change-of-scale problems. In view of the quadratic quantile tests' simplicity, ease of implementation and real-time processing qualities, they are particularly attractive in practical applications. Finally, an integral representation is introduced which permits asymptotic analysis and comparison of a large class of robust detection procedures of which the m-interval and rank detectors are special subclasses. A particularly simple form of nonrank permutation detectors is studied in detail and it is shown that these detectors approach the performance of the locally most powerful detector. It is also shown that, in general, the nonrank detectors are more robust, easier to implement and slightly less powerful than the competitive rank detectors. In conclusion, new directions for research in the area of partition detectors are delineated.

II. PARTITION METHODS: NONPARAMETRIC METHODS BASED ON m-INTERVAL STATISTICS -- FORMULATION AND PROPERTIES

A. Introductory Remarks

It is well known that the problem of detecting signals imbedded in noise may be considered as that of hypotheses testing. For the purpose of this and remaining sections of the chapter, the received observables are represented by either an output of a sampling gate or some other discrete random variables upon which the hypotheses testing is performed. Furthermore, it is assumed throughout the chapter that the components of the noise vector are independent, identically distributed random variables (i.i.d.r.v.) with an absolutely continuous c.d.f. In nonparametric detection problems, the distribution of the observables cannot be completely specified by a finite set of parameters. Based on some general properties of the unknown distributions, detection procedures are devised with two basic properties: robustness and distribution-free. Robustness in this context is a qualitative description which refers to the relative insensitivity of the detector to variations of the underlying noise distribution. It is the most frequent usage to measure the robustness on the basis of ARE in the Pitman-Noether sense. The distribution-free property of the detector is the fixed performance (usually Type I error) of the detector when the null signal (no signal) is transmitted. One may view the latter property as a form of robustness.

The design of actual nonparametric detectors may be considered as a two-stage process. First, the n-dimensional sample space is partitioned into a finite number of regions based only on a few general properties of the noise distributions. These are known as <u>nonparametric partitions</u>. In the second stage of the detector, a set of test statistics is defined as a function of the preselected nonparametric partitions. In this sense every known nonparametric detector is a partition detector.

A commonly used class of nonparametric partitions is generated by the set of hyperplanes $z_i = z_j$, for all $i \neq j$, resulting in the class of rank detectors. Another useful partitioning of the sample space may be generated by the set of hyperplanes, $z_k = a_i$, $i - 1,2,\ldots,m-1$, $k = 1,2,\ldots,n$ where a_i is a set of predetermined constants, resulting in the m-interval partitioning. Most of this chapter will be devoted to the study of m-interval detectors. Particularly, their robustness, simplicity of implementation, flexibility and relatively high efficiency will be explored. The rank detectors will be only introduced as a means of comparison with other partition detectors.

Since the nonparametric partitioning is based only on some general properties of the noise and since the test statistics as set functions of the partitions perform an averaging process, the relative insensitivity to changes in noise distribution is guaranteed. If the partitioning is performed with the signal absent, the probability of receiving an observable in each of the nonparametric partitions is the same or the distribution-free property of detectors is assured. We may compare the nonparametric and parametric detection as follows. In parametric detection, the distance measure and hence the test statistic may be defined for every two points in the sample space. In nonparametric detection, the distance is by contrast defined between a point and a nonparametric partition.

B. Formulation of a Linear m-Interval Detector (LMID)

The LMID requires the specification of m quantiles $\{a_1,\ldots,a_m\}$, where $a_i = F^{-1}(i/m)$ of the c.d.f. $F(z)$. If the quantiles are not known, they may be easily estimated by an efficient recursive procedure.

The quantiles are used to partition the sample space into m+1 disjoint regions $A_i = (a_{i-1}, a_i]$, $i = 1,\ldots,m$ where $a_o = -\infty$, $a_m = \infty$ and $f(a,o) = f(a_m) = 0$, where $f(\cdot)$ is the p.d.f. associated with the c.d.f. $F(\cdot)$.

Consider a sequence of i.i.d. samples of additive noise $\{Z_i, 1 \leq i \leq L\}$ and denote the corresponding signal values $\{s_i, i = 1,\ldots,L\}$, where $s_i = s(t_i)$, $t_1 < t_2 < \ldots < t_L$. The detection problem is then in the form of statistical test of hypotheses:

$$H_o = \prod_{i=1}^{L} F(Z_i) \text{ ; No signal present, i.e., } s(t_i) = 0, i = 1,\ldots,L$$

$$H_1 = \prod_{i=1}^{L} F(Z_i - \theta s_i) \text{ ; Signal present, i.e., } s_i = s(t_i),\ i = t,\ldots,L \tag{1}$$

where θ is a multiplicative factor used to adjust the amplitude signal-to-noise ratio. The LMID is a set function defined on the partitioned space of received signal $x_k = s_k + Z_k$ in the form

$$T = \frac{1}{L} \sum_{k=1}^{L} \sum_{i=1}^{m} b_{ki} I_{A_i}(x_k) \tag{2}$$

where $I_A(\cdot)$ is the indicator function of the set A and b_{ki} are referred to as <u>scores</u>. The test statistic of equation (2) is distribution-free in the sense that its Type I error is independent of the noise distribution, which follows from the fact that under H_o, $\text{Prob}(x_k \in A_i) = F(F^{-1}(a_i) - F(F^{-1}(a_{i-1})) = 1/m$, since T is a linear combination of the random variables $I_{A_i}(x_k)$ each of which is independent of the form of the noise distribution, T itself is distribution free. Each received sample x_k affects T only through the partition A_i into which it falls and thus the incoming samples x_k are effectively quantized. Analysis of the LMID for time-varying signals under the alternative (hypothesis H_1) is considerably complicated by the fact that the samples are independent but not identically distributed. The assumption is made that the scores $b_{ki} =$

$d_k c_i$, i.e., the effects of the signal and the partition act independently upon the scores. With this assumption, the mean and variance of the LMID under H_o can be calculated. Letting

$$\bar{d} = \frac{1}{L} \sum_{k=1}^{L} d_k$$

$$\bar{c} = \frac{1}{m} \sum_{i=1}^{m} c_i$$

$$\overline{d^2} = \frac{1}{L} \sum_{k=1}^{L} d_k^2$$

and

$$\overline{c^2} = \frac{1}{m} \sum_{i=1}^{m} c_i^2$$

the expressions for the mean and variance are

$$E[T/H_o] = \bar{d}\,\bar{c} \tag{3}$$

and

$$\mathrm{Var}\,(T/H_o) = \frac{1}{L}\,(\overline{c^2}\;\overline{d^2} - \bar{c}^2\bar{d}^2) \tag{4}$$

Under H_1, the expected value of T is given by

$$E[T/H_1] = \frac{1}{L} \sum_{k=1}^{L} \sum_{i=1}^{m} c_i d_k [F(a_i - \theta s_k) - F(a_{i-1} - \theta s_k)] \tag{5}$$

C. Most Powerful and Locally Most Powerful m-Interval Detectors

Assuming that the noise distribution is known, the optimum set of scores may be obtained by applying the Neyman-Pearson lemma leading to the most powerful m-interval detector. Following the usual procedure, the logarithm of the likelihood ratio is formed to obtain the most powerful detector

$$T_m = \frac{1}{L} \sum_{k=1}^{L} \log\{\mathrm{Prob}[x_k \in A_{n_k}/H_1]/\mathrm{Prob}[x_k \in A_{n_k}/H_o]\} \tag{6}$$

where n_k is the <u>partition</u> <u>number</u> into which the kth sample falls. More precisely, $n_k = \text{card}(x_i : a_{k-1} < x_i \le a_k)$, where $\sum_{k=1}^{m} n_k = L$. Since

$$\text{Prob}[x_k \in A_{n_k}/H_o] = \frac{1}{m}$$

and

$$\text{Prob}[x_k \in A_{n_k}/H_1] = F(a_{n_k} - \theta s_k) - F(a_{n_{k-1}} - \theta s_k)$$

equation (6) reduces to

$$T_m = \frac{1}{L} \sum_{k=1}^{L} \log m[F(a_{n_k} - \theta s_k) - F(a_{n_{k-1}} - \theta s_k)] \tag{7}$$

To obtain the locally most powerful (LMP) detector, θ is assumed to be small and the term in the brackets in equation (7) is expanded about $\theta = 0$ into Maclaurin's series. After neglecting second and higher order terms, the resulting LMP detector is based on

$$T_{LMP} = \frac{m}{L} \sum_{k=1}^{L} s_k[f(a_{n_{k-1}}) - f(a_{n_k})] \tag{8}$$

To achieve optimum operation of the locally most powerful detector, the scores b_{ki} should be selected so that $d_k = ms_k$ and $c_i = c_{n_k} = f(a_{n_{k-1}}) - f(a_{n_k})$. For dc signals of unit amplitude, equation (8) reduces to

$$T_{LMP} = \frac{1}{L} \sum_{i=1}^{L} b_i n_i \tag{9}$$

where the scores are $b_i = f(a_{i-1}) - f(a_i)$.

To evaluate the performance of the LMID, it is convenient to use asymptotic relative efficiency. Since T is a sum of independent bounded random variables, it is asymptotically normal under both H_o and H_1. As a result of this fact, the ARE may be calculated using the ratio of efficacies. For the LMID an expression for the efficacy follows from its definition and equations (4) and (5)

$$\mathcal{E} = \lim_{L\to\infty} (\frac{\partial}{\partial\theta} E[T/H_1])^2 / L \, Var(T/H_o)$$

$$= \left\{ \frac{1}{L} \sum_{k=1}^{L} \sum_{i=1}^{m} c_i d_k s_k [f(a_{i-1}) - f(a_i)] \right\}^2 / \overline{d^2} (\overline{c^2} - \bar{c}^2) \qquad (10)$$

For the locally most powerful detector with optimum scores, equation (10) reduces to

$$\mathcal{E}(T_{LMP}) = \left(\frac{1}{L} \sum_{k=1}^{L} s_k^2 \right) m \sum_{n_k=1}^{m} [f(a_{n_{k-1}}) - f(a_{n_k})]^2 \qquad (11)$$

If the signals are of unit average power, equation (11) simplifies to the result obtained by Ching and Kurz [1] .

Consider equiprobable partitioning, i.e., $F(a_i) - F(a_{i-1}) = 1/m$ (it will be shown in section III that this partitioning retains the maximum amount of information about the original random variable), and constant level signaling, the efficacy of equation (11) reduces to

$$\mathcal{E}(T) = m \sum_{i=1}^{m} [f(a_{i-1}) - f(a_i)]^2 \stackrel{\Delta}{=} I_m(f) \qquad (12)$$

Taking the limit of $I_m(f)$ as $m \to \infty$ results in

$$\lim_{m\to\infty} I_m(f) = \lim_{m\to\infty} \sum_{i=1}^{m} \frac{\{[f(a_i) - f(a_{i-1})]/(a_i - a_{i-1})\}^2}{\frac{1}{m}(\frac{1}{a_i - a_{i-1}})} (a_i - a_{i-1})$$

$$= \int_{-\infty}^{\infty} \left[\frac{f'(x)}{f(x)} \right]^2 f(x)dx = E\{[\frac{\partial}{\partial\Delta} \ell n f(x - \Delta)]^2\} \Big|_{\Delta = 0}$$

$$= I(f; \Delta = 0) \qquad (13)$$

where $I(f)$ is Fisher's information.

Consider the locally most powerful detector of the form

$$T_o = \frac{1}{L} \sum_{k=1}^{L} \ell n \left[\frac{f'(x_k)}{f(x_k)} \right]$$

The efficacy of this detector is

$$\mathcal{E}(T_o) = \int_{-\infty}^{\infty} \left[\frac{f'(x)}{f(x)}\right]^2 f(x)dx = I(f) \tag{14}$$

Since T_o represents the locally most powerful detector, its efficacy is the largest among all the detectors in the same noise environment, or it follows that the efficacy of the m-interval detector is bounded by Fisher's information. For most noise distributions $m = 6$ results in the efficacy exceeding 90% of its maximum value or LMID have high ARE. Considering the fact that m-interval detectors may be implemented with simple threshold devices and a counter, and they do not require the expensive ordering of rank procedures, the high efficiency combined with robustness makes their practical application particularly attractive. (This point will be further elaborated upon in the next section.)

D. Concluding Remarks

In this section the basic properties of linear m-interval detectors have been conceptualized. Asymptotic bounds have been obtained in comparing the linear m-interval detectors with locally most powerful detectors through the use of Fisher's information. Though omitted from consideration in this section, one can find the distribution of the LMID for finite m and n in terms of two-dimensional generating functions and asymptotic performance for moderate signals in terms of Bahadur and Hodges-Lehmann efficiencies (see reference [1]). Special forms, optimization and application of LMID are discussed in the next section.

III. OPTIMIZATION AND APPLICATION OF LINEAR m-INTERVAL DETECTORS

A. Introductory Remarks

In the first part of this section simple to implement binary threshold and M-ary "matched-filter" type LMID are introduced and evaluated on the basis of ARE. For M-ary systems this requires the introduction of a difference statistic with the associated efficacies, cross-efficacies and correlation coefficient.

In the second part, the equivalence of the LMID and the likelihood ratio detector (LRD) as the number of partitions approaches infinity is shown. An expression for the error probability is derived and a tight upper bound on this probability is presented. Further, the optimality properties of LMID under probability of error and maximum mutual information between the data samples and the partition m-tuples are investigated.

Finally, the robustness and practical usefulness of the LMID is determined by designing both the LMID and LRD for Gaussian noise, but then evaluating the error probabilities for a mixture of Gaussian and impulsive noise. The efficiency and robustness of the LMID in the latter environment is determined by optimizing its design parameters and then computing and comparing its error performance to that of the optimal detector.

B. Two Generalized Sign Detectors

A particularly simple to implement LMID which generalizes the sign detector is based on

$$T = \frac{1}{L}\sum_{i=1}^{m} in_i = \frac{1}{L}\sum_{j=1}^{L}\sum_{i=1}^{m} u(x_j - a_i) \tag{15}$$

where $u(\cdot)$ is the unit step function.

For this detector the efficacy for shift-of-the-mean alternatives reduces to

$$\mathcal{E}_{S.M.} = \frac{\{\sum_{i-1}^{m} i[f(a_{i-1}) - f(a_i)]\}^2}{\sum_{i=1}^{m} i^2[F(a_i) - F(a_{i-1})] - \{\sum_{i=1}^{m} i[F(a_i) - F(a_{i-1})]^2\}} \tag{16}$$

in which the vector of quantiles $\vec{a} = (a_1,\ldots,a_{m-1})$ is as yet unspecified.

Two useful choices of $\vec{a}$ result in two detectors: (1) The choice of partitions is equiprobable or $F(a_i) - F(a_{i-1}) = 1/m$ yielding a generalized-sign equiprobable partitioning detector (GSEPD). (2) The choice of the quantile vector $\vec{a}$ is such that the mean of the generalized sign statistic is maximized or $F(a_i) - F(a_{i-1}) = 2i/[m(m+1)]$ and $F(a_i) = [i(i+1)]/[m(m+1)]$ yielding a generalized-sign maximimum-mean partitioning detector (GSMMPD), which is particularly useful in M-ary transmission.

For the GSEPD equation (16) reduces to

$$\mathcal{E}_{S.M.(1)} = \frac{12}{m^2 - 1}\{\sum_{i=1}^{m} i[f(a_{i-1}) - f(a_i)]\}^2 \qquad (17)$$

If $m = 2$ and $f(a_o) = f(a_2) = 0$, GSEPD becomes a sign detector and $\mathcal{E}_{S.M.(1)} = 4f^2(a_1)$ -- a well-known result.

For the GSMMPD equation (16) reduces to

$$\mathcal{E}_{S.M.(2)} = \frac{18}{m^2 + m - 2}\{\sum_{i=1}^{m} i[f(a_{i-1}) - f(a_i)]\}^2 \qquad (18)$$

Numerical calculations have been performed for both detectors under Gaussian noise conditions. The ARE of both m-interval detectors are over 90% for m = 5, with GSMMPD having a slight edge over the GSEPD. The ARE in non-Gaussian noise with respect to the locally most powerful detector for Gaussian noise frequently exceeds unity. As compared to rank detectors both detectors have slightly reduced efficiency in Gaussian noise but improved robustness if noise distributions vary. In addition, the two detectors are simple to implement and do not require expensive ranking of the observables.

C. Generalized Sign M-ary Detection

Unlike in the binary detection problem, a single partitioning based on the noise distribution is replaced by M partitionings, one "matched" to each hypothesis. For each hypothesis we must pick a partitioning vector $\vec{a}_i$ and a statistic T_i. Thus, for LMID

$$T_i = \frac{1}{L} \sum_{k=1}^{m} b_k n_{ik} \tag{19}$$

where the weight vector $\vec{b}$ of positive components is specified. An intuitively appealing choice of $\vec{a}_i$ is such that

$$E[T_i/H_i] = \vec{b} \cdot \vec{p}_i \tag{20}$$

are maximized, where $\vec{p}_i = (p_{i,1},\ldots,p_{i,m})$ and $p_{i,j} = \text{Prob}[a_{i,j-1} < x \leq a_{i,j}]$. Since

$$\sum_{j=1}^{m} p_{i,j} = 1$$

the optimum $\vec{a}_i$ satisfy

$$F_i(a_{i,j}) - F_i(a_{i,j-1}) = \frac{b_j}{\sum_{k=1}^{m} b_k}, \quad j = 1,\ldots,m \tag{21}$$

Since $F_i(a_i,0) = 0$, by iteration,

$$F_i(a_{i,j}) = \frac{\left[\sum_{k=1}^{j} b_k\right]}{\left[\sum_{k=1}^{m} b_k\right]} \tag{22}$$

are obtained.

Specifically, T_i, $i = 1,\ldots,M$ are selected as GSMMP statistics considered in the previous subsection. In this case

$$T_i = \frac{1}{L} \sum_{k=1}^{m} k n_{ik} \tag{23}$$

where the quantile vectors are given by the solution of

$$F_i(a_{i,k}) = \frac{k(k+1)}{m(m+1)} \qquad i = 1,\ldots,M \quad \text{and} \quad k = 1,\ldots,m \tag{24}$$

When the design of a detector is based on one hypothesis (e.g., the null hypothesis), as is the case in binary detection, the efficacy does not depend on the actual hypothesis separation but only the local rate of separation. This is not any more a meaningful measure of performance when M "matched" detectors are used in an M-ary system. If we assume stochastically ordered hypotheses as representing the M-ary system, the major contribution to the error result from a decision on a hypothesis adjacent to a true hypothesis. Thus we need a measure of performance of two adjacent statistics, T_i and T_j. We replace the relative comparison to the equivalent threshold comparison based on the difference statistic.

$$T = T_i - T_j \tag{25}$$

The efficacy associated with the difference statistic is then a meaningful measure of performance of a "matched" type M-ary detector. The efficacy of T is

$$T = \lim_{L\to\infty} \frac{m_i^2 + m_j^2 - 2m_i m_j}{L(\sigma_i^2 + \sigma_j^2 - 2\rho\sigma_i\sigma_j)} = \lim_{L\to\infty} \left\{ \left[\frac{m_i^2}{L\sigma_i^2} + \frac{m_j^2}{L\sigma_j^2} - \frac{2m_i m_j}{L\sigma_i^2} \right]^{-1} + \left[\frac{m_i^2}{L\sigma_j^2} + \frac{m_j^2}{L\sigma_j^2} - \frac{2m_i m_j}{L\sigma_j^2} \right]^{-1} - 2\rho \left[\frac{m_i^2}{L\sigma_i\sigma_j} + \frac{m_j^2}{L\sigma_i\sigma_j} - \frac{2m_i m_j}{L\sigma_i\sigma_j} \right]^{-1} \right\} \tag{26}$$

where

$$m_k = \frac{\partial}{\partial} E\{T_k\} \Big|_{= 0} \qquad k = i \text{ or } j$$

and

$$\text{var } T = \sigma_i^2 + \sigma_j^2 - 2\rho\sigma_i\sigma_j$$

with $\quad \sigma_k^2 = \text{var } T_k \;, \; k = i \text{ or } j$

$$\rho + E\{(T_j - \bar{T}_i)(T_i - \bar{T}_k)\}/\sigma_i\sigma_j$$

Introducing self- and cross-efficacies

$$\mathcal{E}_{ii} = \lim_{L\to\infty} \frac{m_i^2}{L\sigma_i^2} \; , \; \mathcal{E}_{ij} = \lim_{L\to\infty} \frac{m_i^2}{L\sigma_j^2}$$

the efficacy of the difference statistic is

$$\mathcal{E}_T = [(\mathcal{E}_{ii} - \mathcal{E}_{ji})^{-2} + (\mathcal{E}_{ij} - \mathcal{E}_{jj})^{-2} - 2\rho(\mathcal{E}_{ii}\mathcal{E}_{ij} + \mathcal{E}_{ji}\mathcal{E}_{jj} - \mathcal{E}_{ii}\mathcal{E}_{jj})^{-1}] \tag{27}$$

The self- and cross-efficacies in equation (27) are easy to calculate. The calculation of ρ in specific cases, though laborious, does not create extreme difficulties. The difference efficacy of equation (27) may be used in the same fashion as transition probabilities in evaluation of M-ary m-interval detectors.

D. Comparison of LMID and LRD

As before, let $X = (x_1, x_2, \ldots, x_L)$ be a set of i.i.d. observables used to test the hypothesis H_o with c.d.f. $F(x/H_o)$ against the alternative H_1 with c.d.f. $F(x/H_1)$. Both distributions are continuous and strictly increasing. Let $\vec{a}$ be the vector of constants which partitions the sample space into m-disjoint intervals. The partitions are selected so that the probability of an observable in a particular interval under H_o and H_1 is non-zero. The optimum test statistic based on the m-tuple $N = (n_1, n_2, \ldots, n_m)$, $n_k = \text{card}\{x_i : a_{k-1} < x_i < a_k\}$ is the logarithm of the likelihood ratio

$$T_N = \ell n[P(N/H_1)/P(N/H_o)] \tag{28}$$

The r.v. n_k, $k = 1, 2, \ldots, m$, are multinomially distributed

$$P(N/H_i) = L! \prod_{j=1}^{m} \frac{(p_{ij})^{n_j}}{n_j!} \tag{29}$$

where

$$p_{ij} = F(a_j/H_i) - F(a_{j-1}/H_i) \tag{30}$$

and $i = 0$ or 1 . Alternately, equation (28) may be written as

$$T_N = N \cdot \vec{b} = \sum_{j=1}^{m} n_i b_i \tag{31}$$

where $\vec{b} = (b_1,\ldots,b_m)$ and $b_j = \ell n(p_{ij}/p_{oj})$.

The following theorem indicates expectations for high efficiency of LMID under all signal-to-noise ratios (the analysis in previous sections concentrated only on small signal performance as measured by ARE).

Theorem 1: If under binary hypotheses testing

1) under the hypothesis and the alternative the observables $(x_1,\ldots,x_L)$ are i.i.d.r.v. from a sample space Ω ,

2) the c.d.f.s. under the hypothesis, $F(x/H_o)$, and under the alternative, $F(x/H_1)$, are continuous and strictly increasing,

3) the partitions, $\vec{a}$, are m + 1 quantiles of the joint distribution

$$\vec{a} = \{a_j = H^{-1}(j/m),\ j = 0,1,\ldots,m\}$$

where $H(x) = \alpha F(x/H_o) + (1 - \alpha) F(x/H_1)$ for any $0 < \alpha < 1$, then as $m \to \infty$ the statistic T_N of equation (31) becomes equivalent to the optimum detector which is based on the logarithm of the likelihood ratio.

Proof: For the statistic of equation (31), instead of each interval, consider each random variable, or

$$T_N = \sum_{k=1}^{L} \ell n\left(\frac{p_{1j}(k)}{p_{oj}(k)}\right) \tag{32}$$

where

$$p_{ij}(k) = \Pr\{a_{j-1} < x_k \le a_j/H_i\}\ ,\ i = 0 \text{ or } 1 \tag{33}$$

But $p_{ij}(k) > 0$ and the logarithm is a continuous, strictly increasing function, or

$$\lim_{m\to\infty} T_N = \sum_{k=1}^{L} \ell n \left[\lim_{m\to\infty} \frac{p_{1j}(k)}{p_{oj}(k)} \right] \tag{34}$$

Since (m + 1) quantiles are used as partitions and the distributions under both hypotheses are continuous and strictly increasing, as m increases the interval containing x_k decreases. Thus, by L'Hospital's rule, equation (34) becomes

$$\lim_{m\to\infty} T_N = \sum_{k=1}^{L} \left\{ \ell n \frac{\Pr[x=x_k/H_1]}{\Pr[x=x_k/H_o]} \right\} \tag{35}$$

which is the optimum detector based on the logarithm of the likelihood ratio. Q.E.D.

E. Probability of Error and Bounds on It for the LMID

The decision rule for the LMID is

$N \cdot \vec{b} > K$ select H_1

$N \cdot \vec{b} < K$ select H_o

where K is chosen in accordance with some preselected criterion and ties are resolved by random selection. The expression for the probability of error is then

$$P_e = \sum_{\{N_1\}} P(N/H_o)P(H_o) + \sum_{\{N_2\}} P(N/H_1)P(H_1) + \tfrac{1}{2} \sum_{\{N_3\}} P(N) \tag{36}$$

where $\{N_1\}$, $\{N_2\}$ and $\{N_3\}$ are sets of m-tuples such that $\{N_1\} = \{N : N \cdot \vec{b} > K\}$, $\{N_2\} = \{N : N \cdot \vec{b} < K\}$ and $\{N_3\} = \{N : N \cdot \vec{b} = K\}$, and $P(H_o)$ and $P(H_1)$ are the a priori probabilities associated with each hypothesis.

If the number of data samples L and the number of intervals m are small, the exact probability of error may be calculated because the number of m-tuples is manageable. For a large number of m-tuples, a bound on the probability of error is more realistic to calculate. The following theorem introduces a useful upper bound on the probability of error of a LMID.

Theorem 2: For the LMID defined by equation (31) and if $E[T_N|H_o] \leq K$ and $E[T_N|H_1] \geq K$, the probability of error is bounded by

$$P_e \leq P(H_o) \inf_w \left[\Lambda_o^{-w} \left(\sum_{i=1}^{m} \left(\frac{p_{1i}}{p_{oi}} \right)^w p_{oi} \right)^L \right] + P(H_1) \inf_w \left[\Lambda_o^{-2} \left(\sum_{i=1}^{m} \left(\frac{p_{1i}}{p_{oi}} \right)^w p_{1i} \right)^L \right] \quad (37)$$

where

$$\Lambda_o = \exp K$$

Proof: Starting with equation (36), the probability of error is bounded by

$$P_e \leq P(H_o)\ P[T_N \geq K/H_o] + P(H_1)\ P[T_N \leq K/H_1] \quad (38)$$

with equality satisfied when $P[T_N = K/H_j] = 0$ for $j = 0$ and 1. Applying Chernoff's bound to (38) results in

$$P_e \leq P(H_o) g_T(K/H_o) + P(H_1) g_T(K/H_1) \quad (39)$$

where

$$g_T(K/H_j) = \inf_w [\exp(-Kw) G_T(w/H_j)] \quad (40)$$

and $G_T(w/H_j)$ is the moment-generating function given by

$$G_T(s/H_j) = \left[\sum_{i=1}^{m} \exp(b_i w) p_{ji} \right]^L \quad (41)$$

where

$$b_i = \ln \left(\frac{p_{1i}}{p_{oi}} \right) \quad (42)$$

Substituting from (40), (41), and (42) into (39) results in the desired bound. Q.E.D.

The probability of error expression of equation (36) was evaluated for the shift-of-mean alternatives in Gaussian noise as a function of the signal-to-noise power ratio. Typical curves for $L = 5$ are shown in Figure 1. The probability of error for the LRD is included as a lower bound on the LMID performance. The partitions were selected by a computer gradient search procedure to minimize the probability of error. In obtaining the results, it was assumed that $P(H_o) = P(H_1) = \frac{1}{2}$ and $K = 0$. It is interesting to note that as the number of partitions increases, the performance of the LMID approaches rapidly the performance of the optimum LRD.

F. Maximum Mutual Information Between Data Samples and m-Tuples of Partitioning for the LMID

From a purely intuitive point of view, the partitions which will maximize mutual information between the data samples and m-tuples of partitions should result in a LMID which operates close to its average probability of error. First the relationship between a single data sample and the resulting m-tuple will be deduced and then the results will be generalized to L data samples.

Lemma 1: An m-interval partitioning that is equiprobable retains the maximum amount of information about the original random variable.

Proof:

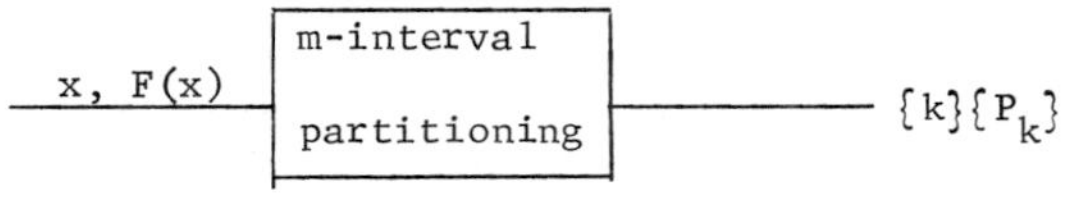

$$H(x) = \int_{-\infty}^{\infty} f(x)\ln[1/f(x)]dx \text{ -- uncertainty about } x$$

$E[H(x/k)] = \sum_{k=1}^{m} H(x/k)P_k$ -- average uncertainty about x after observing the value k

$I(x,\{k\}) = H(X) - H(X/\{k\})$ = decrease in uncertainty about x as a result of observing which interval occurred = information flowing through the m-interval partitioning.

Thus, the probabilities $\{P_k\}$ must be chosen to maximize $I(X,\{k\})$.

But

$$I(X,\{k\}) = \int_{-\infty}^{\infty} f(x)\ell n[1/f(x)]dx - \sum_{k=1}^{m} P_k \int_{a_{k-1}}^{a_k} [f(x)/P_k]\ell n[P_k/f(x)]dx$$

$$= \int_{-\infty}^{\infty} f(x)\ell n[1/f(x)]dx - \sum_{k=1}^{m} \int_{a_{k-1}}^{a_k} f(x)\ell n[1/f(x)]dx - \sum_{k=1}^{m} \cdot$$

$$\cdot \ \ell nP_k \int_{a_{k-1}}^{a_k} f(x)dx$$

Since the first two terms on the right side are equal,

$$I(X,\{k\}) = \sum_{k=1}^{m} P_k \ell nP_k$$

It is well known that the last expression achieves its maximum when $P_k = 1/m$, $k = 1,2,\ldots,m$. Q.E.D.

Theorem 3: The partition $A_i = (a_{i-1}, a_i$, $i=1,2,\ldots,$ m mutual information between the data samples $X = (x_1,x_2,\ldots,x_L)$ and the m-tuples $N = (n_1,n_2,\ldots,n_m)$, maximizes the entropy of the m-tuples $H(N)$.

Proof: The Mutual Information between the data samples and the m-tuple is

$$I(X;N)=E(\log[f(x_1,\ldots,x_n\ N)/f(x_1,\ldots,x_n)])$$

$$= \sum_{\{N\}} \int\ldots\int \log\left[\frac{f_J(x_1,\ldots,x_L,N)}{f_J(x_1,\ldots,x_L)}\right] f_J(x_1,\ldots,x_L,N)dx_1,\ldots dx_L \qquad (43)$$

where $f(x_1,\ldots,x_n,N)$ is the joint probability density function of $(x_1,\ldots,x_n)$ and N. {N} is the set of all possible m-tuples. Then the mutual information may be rewritten as

$$I(X;N) = \sum_{\{N\}} \int\cdots\int f_J(N I x_1,\ldots,x_L)\log\left[\frac{f_J(N\ x_1,\ldots,x_L)}{P(N)}\right] f_J(x_1,\ldots,x_L)dx_1\ldots dx_L. \tag{44}$$

Breaking this integral into n! disjoint regions, each of which contributes equally, one has

$$I(X;N) = \sum_{\{N\}} n! \int\cdots\int_{x_1\ x_2\ \cdots\ x_L} f_J(N\ x_1,\ldots,x_L)\log\left[\frac{f_J(N\ x_1,..,x_L)}{P(N)}\right] f_J(x_1,\ldots,x_L)dx_1\ldots dx_L \tag{45}$$

where $f_J(N\ x_1,\ldots,x_L)=I_{\{x_1,\ldots,x_{n_1}\ A_1\}} x \ldots x I_{\{x_{n_{m-1}+1},\ldots,x_{n_m}\ A_m\}}$.

Substituting this last expression into the integral and integrating each variable out with the knowledge that these are ordered random variables yields, yields,

$$I(X;N) = -\sum_{\{N\}} \frac{L!}{n_1!\ldots n_m!} p_1^{n_1} p_2^{n_2} \cdots p_m^{n_m} \log P(N) \tag{46}$$

Theorem 4: If H(N) is the entropy of the m-tuples and $a_o, a_1,\ldots,a_m$ are the partitions

$$\left.\frac{\partial H(N)}{\partial a_j}\right|_{a_i = F_j^{-1}(k/m)} = 0 \qquad i = 1,2,\ldots,m \tag{47}$$

where $F_j^{-1}(\cdot)$ is the inverse of the joint c.d.f.

Proof: The partial derivative of the entropy is

$$\frac{\partial H(N)}{\partial a_j} = \Sigma P(N)[\ell n P(N) + 1] \frac{n_j}{\Phi_j} - \frac{n_{j+1}}{\Phi_{j+1}} f_J(a)_j \tag{48}$$

where $f_J(\cdot)$ is the joint p.d.f. of a data sample and other terms were defined previously. If the partitions are the joint quantiles, equation (48) reduces to

$$\left.\frac{\partial H(N)}{\partial a_j}\right|_{P_j=1/m} = -\sum_{\{N\}} P(N)[\ell n P(N) + 1]m(n_j - n_{j+1})f_J(a_j) \qquad (49)$$

$$j = 1, \ldots, m$$

The terms in the summation of equation (49) add to zero which is deduced as follows. All m-tuples such that $n_j = n_{j+1}$ yield terms which are identically zero. For every m-tuple such that $n_j - n_{j+1} = k < 0$ there exists an m-tuple such that $n_j - n_{j+1} = k$ formed simply by interchanging the values of n_j and n_{j+1}. Furthermore, since all all the intervals are equally likely, the probabilities of the two m-tuples are equal. Thus, for every positive term in the summation there exists a negative term of equal magnitude, or the right side of equation (49) reduces to zero. Since zero on the right side of equation (49) implies equation (47), the theorem is proven.

The last theorem demonstrates that the joint quantile partitions satisfy the necessary conditions for maximum of the entropy, H(N), or, from equation (46), mutual information, I(X,N). It will now be deduced that this is actually a maximum. Consider that each data sample is transformed into an m-tuple N_i, which has only one nonzero component equal to one. As the next step, the m-tuples are added. It is well known that the entropy of the sum is maximized if the entropy of each component is maximized. In addition, it was shown previously that the entropy of each component is maximized when the partitions correspond to the joint quantiles or an actual maximum is obtained.

To demonstrate the effectiveness of the LMID under this design approach, the joint quantile partitions were computed for the shift-of-mean alternatives in Gaussian noise. The results for $L = 5$ are shown in figure 2.

The curves verify that this approach provides near optimum performance if m is even. The poor performance for $m = 3$ or 5 is due to the fact that the center intervals are inappropriately wide.

G. Performance of the LMID in Noise Represented by a Mixture Distribution Model

The importance of selection of proper detectors to perform well in a noise environment represented by a mixture distribution model has been treated in the literature for the past fifteen years [9-15]. In the first part of this subsection the relative robustness of the LMID and the LRD, where both detectors are designed on the assumption that the noise samples are Gaussian while they are actually a mixture of Gaussian and impulsive noise, is investigated. In the second part, it is shown how the probability of error of the LMID can be minimized for mixture noise (Gaussian and impulsive) by the proper selection of partitions and/or scores.

As suggested in [16], the impulsive noise is represented by the Cauchy distribution. Though any other impulsive noise model with large tails of the p.d.f. may be used with equal success. If comparisons are made between LMID and LRD designed for Gaussian noise when actually Gaussian and impulsive noise are present, the following phenomena are observed. The probability of error of the LRD is plotted as a function of the signal-to-noise ratio for $L = 2$ in figure 3 and $L = 5$ in figure 4, for the indicated impulsive noise conditions. The shaded areas indicate the regions for a family of probability curves for a given value of δ, the rate of occurrence of impulses. It is interesting to note that when the impulsive noise dominates the probability of error, increasing the number of samples from 2 to 5 causes the probability of error to rise. This is due to the fact that almost regardless of the magnitude of the signal, if an impulse occurs in the signaling interval, it dominates the test statistic. The probability of error of the LMID as a function of signal-to-noise ratio is plotted in figures 5 and 6 for $L = 2$ and in figures 7 and 8 for $L = 5$. In figure 7 the curves for $m = 4$ and 5 are omitted since they differ negligibly from the curve for $m = 6$. Similarly, in figure 8 the curve for $m = 5$ is omitted.

In general, the results show that for severe impulsive noise conditions the LMID remains robust while the LRD fails to perform within acceptable bounds. While the robustness of the LRD decreases with an increase in the number of samples, the robustness of the LMID improves as the number of samples increases.

If the partitions of the LMID are optimized using the mixture model the reduction in error rate is greatest for high signal-to-noise ratio, large sample sizes and less frequent noise impulses. In all cases the partitions are more closely grouped than the partitions for the Gaussian noise, though the changes in the partitions are frequently very small.

If the partitions and scores are optimized simultaneously, the influence of the scores on the error rate demonstrates some unusual aspects. For instance, optimum scores for Gaussian noise when $m = 2$ or 3 intervals are linear transformations of the optimum scores for the mixture distribution or, in this case, though the scores change as the noise distribution changes, the decision regions and hence the error rate remain the same.

If the Gaussian and impulsive noise scores are not equivalent to the Gaussian noise scores, further reductions in the error rate are observed.

In conclusion, it is interesting to note the high degree of robustness of the LMID. Namely, not only does the LMID designed for Gaussian noise perform well in Gaussian and impulsive noise, but it performs so well that it is frequently difficult to improve measurably upon the performance using the optimum LMID. Furthermore, because the performance of the LMID depends on the interval probabilities and not on the exact shape of the noise distributions, one would expect the LMID to perform well if other models for impulsive noise distribution are used.

IV. ADAPTIVE m-INTERVAL DETECTORS

A. Introductory Remarks

So far in Sections II and III linear m-interval nonparametric detectors have been conceptualized and studied demonstrating their high efficiency, robustness and ease of implementation. The basic problem of finding the quantiles and the p.d.f. at those quantiles when the underlying distributions are unknown has not been resolved. Although it was shown in Section III that choosing the Gaussian distribution as a means of finding the partitions and scores, which are related to the quantiles and p.d.f. at the quantiles, results in a robust detector performing well even if the noise distribution is actually represented by a mixture model, the solution will usually not be satisfactory if the noise distribution is significantly different from Gaussian. In this section two basic approaches are presented which permit an adaptive operation of the m-interval detector. The first approach involves recursive point estimation of quantiles, p.d.f.'s and their derivative at the quantiles by proper adaptation of the stochastic approximation algorithms [19,20,21,22]. The second approach yields an adaptive m-interval detector involving the use of order statistics.

Finally, additional procedures for application of adaptive m-interval detectors in special situations are discussed.

B. Recursive Point Estimation of Quantiles, Probability Density Functions and Their Derivatives

The method of stochastic approximation has ushered in a whole family of recursive estimation procedures which are particularly suited for digital computer processing and real-time applications. If a specific class of problems need be solved, suitable microprocessors are also useful as means of implementation of recursive estimators. In this subsection a method based on stochastic approximation is presented which permits simultaneous estimation of quantiles, the value of the p.d.f. and its derivative at these quantiles (though the derivatives of the p.d.f. at the quantiles are not needed for m-interval detectors, they are useful in other types of nonparametric detectors).

The Robbins-Monro (R-M) procedure [19] is a recursive estimation procedure for locating the root of a regression function in a manner similar to the Newton-Raphson method for the location of a root of a nonlinear function. The essential recursion equation is

$$x_{n+1} = x_n - (C/n)[Y(x_n;\theta) - \alpha] \tag{50}$$

where $Y(x_n;\theta)$ is a family of random variables with index x_n, and θ is the solution of the equation

$$E[Y(x_n;\theta)] - \alpha = M(x_n) - \alpha = 0 \tag{51}$$

Under suitable conditions upon the regression function $M(x_n)$ and the family of random variables $Y(x_n;\theta)$ it has been shown that $x_n \xrightarrow{m.s.} \theta$ [19] and $x_n \xrightarrow{a.s.} \theta$ [20]. In addition, if one denotes the slope of the regression function at $x = \theta$ as α_1 and the $\text{var}[Y(x_n;\theta)] = \sigma^2$, Sacks [21] has shown that $n^{1/2}(x_n - \theta)$ converges in law to $N(0,C^2\sigma^2/(2C\alpha_1 - 1))$.

If $Y(x_n;\theta) = \text{sgn}[x_n - y_n] + 1 - 2a$, where y_n are i.i.d. samples from a population with a c.d.f. $F(x)$, then $M(x) = 2[F(x) - a]$ and x_n converges to the ath quantile of the c.d.f.

The R-M procedure may be also used as a vehicle to obtain estimates of both the p.d.f. and its derivative at the quantiles with little additional cost in storage or computation time. Using a variant of Burkholder's result [22], it is possible to achieve the goal at a cost of two additional samples at each iteration of the R-M procedure. In particular, let the assumptions of R-M procedure hold and, in addition, let $F(x)$, $f(x)$ and $f'(x)$ be continuous in the neighborhood of θ. Choose $\{c_n\}$ to be a positive null sequence satisfying the constraint that $\sum_n (1/n\, c_n^2)^2 < \infty$. Denoting $Y_{3n+i} = u(x_n + ic_n - y_{3n+1})$, $i = -1,0,1$, where $u(\cdot)$ is the unit step function, then

$$\frac{1}{n} \sum_{j=1}^{n} \frac{Y_{3j+1} - Y_{3j-1}}{2c_j} \overset{a.s.}{\to} f(\theta) \tag{52}$$

and

$$\frac{1}{n} \sum_{j=1}^{n} \frac{Y_{3j+1} - 2Y_{3j} + Y_{3j-1}}{4c_j^2} \overset{a.s.}{\to} f'(\theta) \tag{53}$$

The constraint that $\sum_n (1/[nc_n^2])^2 < \infty$ may be relaxed, if one is not interested in obtaining an estimate of $f'(\theta)$, to $\sum_n (1/[nc_n])^2 < \infty$ or Burkholder's Theorem 5 [22] may be applied to obtain an estimate of $f(\theta)$ requiring only one additional sample at each iteration of the R-M procedure. Extensive simulation studies performed on the estimation procedures outlined above lead one to conclude that they are accurate and practicable.

An immediate application of the results to the estimation of optimum scores and partitions of the LMID is obvious. This particular detector requires both the quantiles and the value of the p.d.f. at the quantiles to achieve its optimum performance under low signal-to-noise conditions (locally most powerful detection). The above procedure is ideally suited to supply the required information and update it periodically using a learning interval or a parallel processor. Thus, means for efficient adaptive operation of the LMID are readily available. The R-M procedure achieves its optimal convergence rate when the gain coefficient C of equation (50) is equal to $1/\alpha_1$. In this particular application of the R-M procedure $\alpha_1 = f(\theta)$ and, therefore, one may use the current estimate of $f(\theta)$, after a sufficient number of iterations, directly in equation (50) to improve the rate of convergence of the procedure.

The above method may be easily extended to obtain m distinct quantiles $\theta^{(i)}$ simultaneously with the corresponding values $f(\theta^{(i)})$ and $f'(\theta^{(i)})$, $i = 1,2,\ldots,m$. Moreover, no additional samples are required to obtain this information at the m quantiles because the recursion equation need only be rewritten as

$$x_{n+1}^{(i)} = x_n^{(i)} - \frac{C^{(i)}}{n}\left[\operatorname{sgn}(x_n^{(i)} - y_n) + 1 - 2a^{(i)}\right] \tag{54}$$

for $i = 1,2,\ldots,m$. Note that the same sample y_n is used in each of the m recursion equations. Similar equations can be written to replace equations (52) and (53) for the m distinct quantiles. In short, the R-M procedure in its proper form provides an efficient recursive method for estimating quantiles of a c.d.f., the p.d.f. and its derivative at the quantiles making possible the development of many forms of adaptive nonparametric detectors.

C. Adaptive LMID Using Order Statistics

Consider the LMID of Section II for which the parameters $a_i = f^{-1}(1/m)$, $i = 1,2,\ldots,(m)$ and $P_i = \text{Prob}[z \in A_i]$, $i = 1,2,\ldots,m$ are unknown. It is desired to estimate these parameters using a set of finite-sample-size observables. Thus, at the receiver we have a set of observations from pure noise

$$W = \{w_r,\ r = 1,2,\ldots,N_w\} \tag{55}$$

and a set of observations from a signal and noise sample

$$X = \{x_\nu, \nu = 1,2,\ldots,N_x\} \tag{56}$$

Consider

$$Z_k = \{z_k, k = 1,2,\ldots,n\} \tag{57}$$

an unknown sample to be tested, let $N_w = md - 1$, where d is an integer, and let $w^{(j)}$ be the jth order statistic of noise, then $w^{(id)}$ is an estimate of the quantile a_i. Denote n_i and h_i as the numbers of z's and x's in the interval $(w^{(d(i-1))}, w^{(id)})$, respectively, then an adaptive most powerful m-interval detector takes the form

$$T_a = \sum_{i=1}^{m} \left(\frac{n_i}{n}\right) \ell n \left(\frac{h_i}{N_x}\right) \tag{58}$$

For large N_x and N_w, $w^{(id)}$ converges to a_i and h_i/N_x converges to P_i and T_a of equation (58) reduces to the most powerful m-interval detector. The estimators of a_i and P_i given by $w^{(id)}$ and h_i/N_x, respectively, converge in any noise distribution.

If N_w is assumed to be large, which permits the replacement of $w^{(id)}$ by a_i, the distribution of T_a can be found exactly (see Ching and Kurz [1]).

D. Other Forms of Adaptive m-Interval Detectors

There are numerous forms of adaptive m-interval detectors that are useful in particular applications. For instance, if the underlying noise is of the mixture distribution type, an adaptive m-interval detector may be realized by using the Cohen-Kurz model [10] and replacing the rank detectors by two suitable LMID, one matched to Gaussian the other to impulsive noise. Numerous adaptive m-interval detectors may be generated by using the philosophy of Chadwick and Kurz [18].

Many other approaches or combinations of approaches may be used with considerable success. Intimate knowledge of a particular problem, ingenuity and imagination yield excellent solutions in a given application.

V. NONLINEAR PARTITION DETECTORS

A. Introductory Remarks

Linear m-interval detectors investigated so far in this chapter perform well if the detection problem is of the stochastic ordering type, particularly shift of the mean type. Many practical problems such as detection of signals transmitted through dispersive media and detection of texture edtes in picture processing are better represented by a change of scale model. In the latter case it is intuitively appealing to consider nonlinear partition detectors hoping to achieve high discrimination in change-of-scale detection problems. In this section, the properties of three-sample quadratic quantile statistics are investigated. The performance of these detectors is comparable to that of the three-sample rank detectors but requires a much simpler detection process at the receiver and the test statistics can be formed and updated in real time. The three-sample quadratic quantile detectors exhibit good performance characteristics in change-of-scale problems, but also perform well in stochastic ordering problems thereby yielding an added degree of robustness.

B. Detector Models for Small Signal Detection

Consider the three-sample binary detection problem as conceptualized for rank detectors in references [23] and [24]. Thus it is assumed that the detector has prior knowledge as a result of the acquisition of n-dimensional vectors $\vec{Y}$ and $\vec{Z}$ under the hypothesis (H_o) and the alternative (H_1), respectively, during the learning interval of operation. The vector $\vec{Y}$ consists of i.i.d.r.v. from the c.d.f. $F_y(y)$ and the vector $\vec{Z}$ consists of i.i.d.r.v. from the c.d.f. $F_z(z)$. Let the space of observables be partitioned into $m+1$ intervals subject to

$$\sum_{i=1}^{m} p_i = 1$$

such that

$$\text{Prob}[a_{i-1} < y_j \leq a_i] = F_y(a_i) - F_y(a_{i-1}) = p_i \tag{59}$$

$$\text{Prob}[\alpha_{i-1} < z_j \leq \alpha_i] = F_z(\alpha_i) - F_z(\alpha_{i-1}) = p_i \tag{60}$$

for $i = 0,1,2,\ldots,m$; $j = 1,2,3,\ldots,n$, while a_i and α_i represent the ith quantile of $F_y(y)$ and $F_z(z)$, respectively. With the quantiles so specified, an n-dimensional data vector $\vec{X}$ of i.i.d.r.v. from the c.d.f. $F_x(x)$ is presented to the detector which calculates the random vectors $\vec{h}$ and $\vec{k}$ with components $\{h_i\}$ and $\{k_i\}$ corresponding to the number of x_i values which fall in the intervals $(a_{i-1}, a_i]$ and $(\alpha_{i-1}, \alpha_i]$, respectively. The quantiles needed for the quadratic quantile detection can be easily obtained by the methods of the previous section or similar type of procedures. The quadratic detector is based then on the test statistic

$$T = (\vec{h} - \vec{N})^T \underline{b}(\vec{h} - \vec{N}) - (\vec{k} - \vec{N})^T \underline{c}(\vec{k} - \vec{N}) \tag{61}$$

where $\vec{N} = E[\vec{h}/H_o]$, while $\underline{b}$ and $\underline{c}$ are diagonal matrices of positive elements apportioned to emphasize those quantities which correspond to the regions of greatest dissimilarity between $F_y(y)$ and $F_z(z)$ if the latter information is available. If no information about regions of dissimilarity is available, the matrices $\underline{b}$ and $\underline{c}$ are replaced by identity matrices.

If the small signal model is assumed, a usual assumption for nonparametric detectors, the test statistic of equation (61) may be replaced by

$$T = [(\vec{h} - \vec{N}) - \vec{\mu}]^T \underline{A}[(\vec{h} - \vec{N}) - \vec{\mu}] - \gamma \tag{62}$$

where

$$\underline{A} = \underline{b} - \underline{F}^T \underline{c}\, \underline{F}$$

$$\vec{\mu} = \underline{A}^{-1} \underline{F}^T \underline{c}(\underline{F}\vec{N} - \vec{N}) \tag{63}$$

$$\gamma = [\underline{F}^T \underline{c}(\underline{F}\vec{N} - \vec{N})]^T \underline{A}^{-1} [\underline{F}^T \underline{c}(\underline{F}\vec{N} - \vec{N})] + (\underline{F}\vec{N} - \vec{N})^T \underline{c}[\underline{F}\vec{N} - \vec{N}]$$

Since $\vec{h}$ and $\vec{k}$ are derived from the same input data $\vec{X}$, they cannot be statistically independent. To deal with this dependence the approach can be followed which, although approximate has the advantage of generality, postulates a linear dependence model for large n and m. (Extended simulation studies under small signal conditions for various classes of distributions lead to the conclusion that this is an appropriate approximation.) It is assumed that there is a sufficient number of data points n so that under H_o the location distribution of the h_i points in each $(a_{i-1}, a_i]$ interval is uniform. In essence this allows replacement of $F_y(y)$ by a piece-wise linear model. This approximation is in error in the extreme tail region of the c.d.f., since an asymptotic analysis is the final objective, m can be made as large as necessary to achieve an acceptable approximation error.

The elements of the matrix $\underline{F}$ in equations (63) are given by

$$f_{ij} = \frac{(a_{j-1}, a_j] \cap (\alpha_{i-1}, \alpha_i]}{(a_j - a_{j-1}]} \qquad 0 \le f_{ij} \le 1 \qquad i,j = 1,\ldots,m \tag{64}$$

Thus, there are approximately $h_j f_{ij}$ data points from quantile interval $(a_{j-1}, a_j]$ within the interval $(\alpha_{i-1}, \alpha_i]$ so that

$$k_i = \sum_{j=1}^{m} f_{ij} h_j$$

or $\vec{k} = \underline{F}\vec{h}$.

For shift-of-mean $(\vec{Z} = \vec{Y} + \delta \,,\ \delta \ll 1)$

$$\underline{F} = \begin{bmatrix} 1 & \delta/q_2 & 0 & \cdots & \cdots & 0 \\ 0 & 1 - \delta/q_2 & \delta/q_3 & & & \\ 0 & 0 & 1 - \delta/q_3 & 0 & & \\ \vdots & & & \delta/q_{m-1} & & \\ & & 0 & 1 - \delta/q_{m-1} & & \\ 0 & & & & & 1 \end{bmatrix}$$

where $q_i = (a_i - a_{i-1}]$

and

$$\underline{A} \approx \delta \begin{bmatrix} 0 & \frac{-1}{q_2} & 0 & \cdots & & 0 \\ \frac{-1}{q_2} & \frac{2}{q_2} & \frac{-1}{q_3} & & & \vdots \\ 0 & \frac{-1}{q_3} & \frac{2}{q_3} & & 0 & \vdots \\ \vdots & & & & \frac{-1}{q_{m-1}} & 0 \\ \vdots & 0 & & & \frac{2}{q_{m-1}} & 0 \\ 0 & & & & 0 & 0 \end{bmatrix}$$

For change-of-scale $(F_z(z) = F_y(y/\theta)$

$$\underline{F} = \begin{bmatrix} \underline{F}_1 & \underline{0} \\ \underline{F} & \underline{F}_2 \end{bmatrix} \qquad \underline{A} = \begin{bmatrix} \underline{A}_1 & \underline{0} \\ \underline{0} & \underline{A}_2 \end{bmatrix}$$

where

$$
\underline{F}_1 = \begin{bmatrix}
1 & \gamma\Delta_1 & & & \\
 & 1 - \gamma\Delta_1 & \gamma\Delta_2 & & 0 \\
 & & 1 - \gamma\Delta_2 & \ddots & \\
 & 0 & & \ddots & \gamma\Delta_{\frac{m}{2} - 1} \\
 & & & & 1 - \gamma\Delta_{\frac{m}{2} - 1}
\end{bmatrix}
$$

$$
\underline{F}_2 = \begin{bmatrix}
1 - \gamma\Delta_{\frac{m}{2} + 1} & & & 0 \\
\gamma\Delta_{\frac{m}{2} + 1} & 1 - \gamma_{\frac{m}{2} + 2} & & \\
 & \gamma\Delta_{\frac{m}{2} + 2} & \ddots & \\
 & & \gamma\Delta_{m - 1} & 1
\end{bmatrix}
$$

$$
\underline{A}_1 = \gamma \begin{bmatrix}
0 & -\Delta_1 & & \\
-\Delta & 2\Delta_1 & -\Delta_2 & 0 \\
 & -\Delta_2 & \ddots & \ddots \\
 & & \ddots & -\Delta_{\frac{m}{2} - 1} \\
 & 0 & \ddots & 2\Delta_{\frac{m}{2} - 1} \\
 & & & -\Delta_{\frac{m}{2} - 1}
\end{bmatrix}
$$

$\underline{A}_2$ has similar form involving $\Delta_{(m/2) + 1}$ to Δ_{m-1}, and

$$
\Delta_i \triangleq \begin{cases}
\dfrac{1}{1 - \left(\dfrac{a_{i+1}}{a_i}\right)} & \text{for } i = 1,2,\ldots,(m/2) + 1 \\[2ex]
0 & \text{for } i = m/2,\ m \\[2ex]
\dfrac{1}{1 - \left(\dfrac{a_{i-1}}{a_i}\right)} & \text{for } i = (m/2) + 1,\ldots,m - 1
\end{cases}
$$

$$\gamma a_i \approx [a_i - \alpha_i]$$

C. Asymptotic Normality of the T Statistic

The proofs of asymptotic normality of the two types of quadratic detectors are too lengthy and complicated to be given here (see reference [26] for details). The proofs are based on the application of Whittle's theorem [27]. The results are summarized in Table 1 where $\underline{B}$ is generated similarly to $\underline{A}$ by replacing $\underline{F}$ with $\underline{G}$ and the elements of $\underline{G}$ are given by

$$g_{ij} = \frac{(\alpha_i - \alpha_{j-1}) \cap (a_i - a_{i-1})}{(\alpha_i - \alpha_{j-1})} \tag{65}$$

In addition, ∇_i has the same meaning in $\underline{G}$ and $\underline{B}$ as Δ_i in $\underline{F}$ and $\underline{A}$, $\vec{g} = 1/g\ (\vec{h} - \vec{N})$, $\mathrm{tr}(B^2) = 6\psi^2 \sum_{i=1}^{m} \nabla^2$ and $c = m/n$ (constant).

D. Comparison of Three-Sample Quadratic Quantile and Linear Rank Detectors

The comparison will be given here on the basis of asymptotic average probability of error P_e because only for this measure of performance specific results are available for three-sample linear rank detectors [23]. Using the results given in Table 1 for quadratic quantile detectors and in reference [23] for the linear rank detectors, if shift-of-mean model and small signals ($\delta \ll 1$) are assumed

$$\frac{P_e}{\hat{P}_e} = 1 + \delta\left[\tfrac{1}{2}\sqrt{3/n} - \frac{m^3/2}{\sqrt{2n}\ n}\right] \tag{66}$$

where P_e refers to the quantile detector and $\hat{P}_e$ to the rank detector. As seen from Figure 9, if m is allowed to grow at the same rate as $n([n/m] = c)$, the quadratic quantile detector will perform better than the linear rank detector for $m > \sqrt{3c\pi/2}$. Figure 10 gives the boundary of equal performance. The true significance of the result is that one does not sacrifice performance by substituting the quadratic quantile detector for the three-sample rank detector in shift-of-mean detection problems. Yet the quantile detector is much simpler to implement and is applicable in real-time processing. Furthermore, while the comparison is presented only for the shift-of-mean model, the quantile detector offers a considerable advantage over linear rank detectors for change-of-scale detection problems because of its quadratic form.

VI. UNIFIED APPROACH TO PARTITION DETECTION

A. Introductory Remarks

In this section analytical results, which allow comparisons of a broad class of partition detectors be they of rank or quantile type within a unified framework, are presented. This framework is useful because it permits the comparison of a broad diversity of approaches by the same analytical tools and creates a better understanding of all types of possible nonparametric tests which are neither clearly quantile or rank. In addition, the representation clarifies the relationship between rank and quantile tests and permits the generalization of the latter tests. A particularly simple subset of the general representation is shown to yield good performance while simplifying the implementation in some cases. The primary analytical results of this section are achieved by extending the asymptotic normality theorems of Pyke and Shorack [31] to encompass a broader class of statistics.

B. Representation for the Generalized Linear Test

Consider n_i i.i.d. samples from the distributions F_i , $i = 1,2,\ldots,M$. Form the empirical distribution functions

$$F_{in_i}(x) = \frac{1}{n_i} \sum_{j=1}^{n_i} u(x - x_{ij}) \tag{67}$$

where $u(\cdot)$ is the unit step function (continuous from the right) and x_{ij} is the jth sample from F_i. Define the weighted sum of empirical distribution functions

$$H_N(x) = \sum_{i=1}^{M} \beta_i F_{in_i}(x)$$

where

$$\sum_{i=1}^{M} \beta_i = 1 , \quad \beta_i \geq 0 , \quad \sum_{i=1}^{M} n_i = N$$

The M-dimensional statistic, spanning an (M-1)-dimensional space $\vec{T} = (T_1,\ldots,T_M)$, where

$$T_i = \int_0^1 F_{in_i}(H_N^{-1}(\alpha))d\nu(\alpha) \tag{68}$$

and ν is a bounded measure on [0,1] , is a general representation for a broad class of nonparametric tests. This representation was originally proposed by Pyke and Shorack [31] to develop proofs of asymptotic normality of linear rank tests. An extension of the class of tests for which this representation is valid as well as the proof of asymptotic normality for the extended class of tests will be given in this section. Before proceeding with the mathematical development for the general class of nonparametric tests, the equivalence of the representation given by equation (68) to some well-known nonparametric tests will be presented.

C. Equivalence to Linear Rank and Quantile Tests

If ν is a measure which assigns weights $C_{N,k}$ on the points k/N for $1 \leq k \leq N$ and zero elsewhere, then equation (68) reduces to

$$T_i = \sum_{k=1}^{N} C_{N,k} F_{in_i}(H_N^{-1}(k/N)) \tag{69}$$

where $\beta_i = n_i/N$. For the special case $\beta_i = n_i/N$, $H_N(\cdot)$ is the empirical distribution function of the pooled sample and $H_N^{-1}(k/N)$ is the kth order statistic of the pooled sample. The Chernoff-Savage statistics [32]

$$T_i^* = \sum_{k=1}^{N} C_{N,k}^* Z_{Nki} \tag{70}$$

where Z_{Nki} is one if the kth order statistic from the joint ranking is from F_i and zero otherwise, is identical to the statistics of equation (69) when

$$C_{N,k}^* = C_{N,k+1} + \ldots + C_{N,N} \tag{71}$$

The integral representation of linear rank statistics [32]

$$T_i^* = \int_o^1 J(H_N(x))dF_{in_i}(x) \tag{72}$$

is identical to the representation of equation (68) for

$$J(b) - J(a) = \nu((a,b]) \qquad 0 < a < b < 1$$

Consider the linear quantile tests investigated in Sections II through IV. The T_i tests of those sections are then

$$T_i = \sum_{j=1}^{k+1} C_{N,j} \{\text{number of samples from } F_i \text{ in the interval } (a_{j-1},a_j]\} \tag{73}$$

where the quantiles $\vec{a} = (-\infty,a_1,\ldots,a_k,\infty)$ are either a fixed vector of constants or a random vector based on training samples. The equivalence of equations (73) and (69) may be seen by dividing the M sample sets into two groups. One group consists of M_1 sets of unknown samples from $F_1,\ldots,F_{M_1}$ and the other group consists of M_2 sets of training samples from $F_{M_1+1},\ldots,F_M$. Let $\beta_1 = \beta_2 = \beta_{M_1} = 0$ which results in the vector $\vec{a}$, equation (69) is still useful. By letting $n_{M_1+1},\ldots,n_M \rightarrow \infty$, the convergence with probability one of the empirical distribution functions assures the convergence of $\vec{a}$ to a constant. When $n_1,\ldots,n_{M_1} \rightarrow \infty$ at a slower rate, the asymptotic results allow the linear quantile tests to be included within the same mathematical framework as the rank tests.

D. Asymptotic Normality of the T_i Tests

The T_i statistical tests have been studied by numerous researchers. Their asymptotic normality has been demonstrated with different conditions on the score-generating functions and the distributions F_i [31-35]. Usually if strict regularity conditions are imposed on the scores correspondingly weaker conditions may be imposed on the distributions F_i and, conversely, asymptotic normality may be established with very general scores if the distributions are sufficiently smooth. In view of the fact that the interest in this section is to find a general mathematical framework to compare rank, quantile and related nonparametric tests and smoothness of the underlying distributions is not a severe constraint in the majority of practical detection problems, the effort in this section will be concentrated on asymptotic normality of T_i tests with weakly restricted score-generating functions, particularly those which generate step-function scores. Following Pyke and Shorack [31], very weak conditions on the scores may be imposed if the functions $F_i(H^{-1}(\alpha))$ are assumed to be uniformly continuous on $(0,1)$. Though this condition may be weakened somewhat, uniform continuity simplifies the proofs of asymptotic normality.

The original proofs of Pyke and Shorack are inapplicable if the quantile detectors of Sections II and III are to be included. In particular, it is necessary to allow zero values of the weights β_i and to allow the sample sizes n_i to grow at unequal rates. These two new conditions require the proof of the asymptotic normality to be modified and extended. The proofs of theorems of this type use a technique due to Doob [36] of representing the limiting form of the empirical distribution function as a separable tied-down Wiener process. It is then possible to represent the limit of T_i as the integral of a Wiener process. The asymptotic normality theorems require a precise set of conditions imposed on both the score-generating function and the distributions. The new version of the theorem presented here requires the imposition of two additional conditions -- one on the distributions beyond those imposed in [31] and one on the maximum difference of growth rates of the sample sizes.

Assumption 1: The functions $F_i(H^{-1}(\alpha))$ have derivatives which form a uniformly equicontinuous family on (0,1).

Assumption 2: For each $i,j = 1,\ldots,M$ such that $\beta_j \neq 0$, $\beta_i \neq 1$, there exists a constant k such that

$$F_i F_j^{-1}(t) \leq kt \qquad \text{on} \qquad 0 \leq t \leq 1$$

where $F_i F_j^{-1}(t)$ is a shorthand notation for $F_i(F_j^{-1}(t))$. In addition, there exists an $\epsilon > 0$ such that

$$\frac{\partial}{\partial t} F_i F_j^{-1}(t) < \infty \qquad \text{on} \qquad [0,\epsilon]$$

Assumption 3: The set of measures ν_N converges to the measure ν in the sense that

$$\int_{1/n_i}^{1} L_N d(\nu_N - \nu) \underset{P}{\rightarrow} 0$$

where

$$L_N(t) = \sqrt{n_i}\,[F_{in_i} F_i^{-1}(t) - t]$$

Assumption 4: Let D be the set of all right-continuous real-valued functions on [0,1] having only jump discontinuities. Let Q' denote the class of all non-negative functions defined on [0,1] which for some $\epsilon > 0$ are bounded away from zero on $(\epsilon, 1 - \epsilon)$, non-decreasing on $[0,\epsilon]$ and non-increasing on $[1 - \epsilon, 1]$, and which have square-integrable inverses. Let

$Q = \{q \in D : q > q'$ for some $q' \in Q'\}$. Assume

$$\int_0^1 q\,d|\nu| < \infty$$

for some $q \in Q$.

Assumption 5: Let

$$m = \min_{\substack{1 \le i \le M \\ \{i:\beta_i \neq 0\}}} \{n_i\} \quad \text{and} \quad n_{min} = \min_{1 \le i \le M} \{n_i\}$$

assume that

$$\lim_{n_{min} \to \infty} (n_{min}^{\frac{1}{2}}/m) = 0(q^{-1}(1/m))$$

Comments on the Assumptions: Assumptions 1 and 2 restrict the class of distributions. Assumption 2 is new to allow zero values of β_i. Assumptions 3 and 4 are mild constraints on the behavior of the measure ν and they are the same as in [31]. Assumption 5 limits the minimum sample size increases. These are not the only variations on the assumptions by Pyke and Shorack. Other sets of useful constraints are possible. In some practical applications it is convenient to replace assumption 2 by a less restrictive one which will not change the validity of the theorem.

Theorem: Let $\vec{T}$ represent the vector of all T_i for which $\beta_i \neq 1$. Then under the assumptions 1-5, $\vec{T}$ converges in law to a jointly normal distribution with

$$E[T_i] = \int_0^1 F_i H^{-1}(\alpha)\,d\nu(\alpha) \tag{74}$$

and

$$\operatorname{cov}(T_i, T_j) = 2n_{min}^{-1} \sum_{k=1}^{M} \alpha_k \int_0^1 \int_0^\alpha (\delta_{ik} - \beta_k b_i(t))$$

$$\cdot\, (\delta_{jk} - \beta_k b_j(t)) F_k H^{-1}(\beta)[1 - F_k H^{-1}(\alpha)]\,d\nu(\alpha)\,d\nu(\beta) \tag{75}$$

where δ_{ij} is the Kroneckker delta and

$$b_i(\alpha) = \frac{\partial}{\partial\alpha} F_i H^{-1}(\alpha) \ , \ \alpha_k = \lim_{n_{min}\to\infty} \frac{n_{min}}{n_k}$$

Proof: The proof of the theorem is given in the appendix. Though equations (74) and (75) appear to be complicated, it will be shown that considerable simplification is possible in many practical tests for which Theorem in this section is valid.

E. The Generalized Quantile Test

Consider a special class of detectors, for which Theorem in this section is satisfied, with the measure ν chosen so that it assigns weights C_j to the points $0 < a_1 < \ldots < a_m < 1$. Define

$$T_1^* = \sum_{j=1}^{m} C_j W_i(a_j) \tag{76}$$

where

$$W_i(a_j) = F_{in_i} H_N^{-1}(a_j)$$

The test statistic of equation (76) represents a broad class of threshold tests. Depending on the choice of β_i and the relative growth rates of the n_i , equation (76) may be used for asymptotic analysis of quantile and joint quantile tests with random and fixed partitions. As it was indicated before, in partition tests the number of thresholds m is a measure of the robustness of the test. In the next subsection it will be shown that the efficiency of the T_i^* tests is high for small m , indicating that the penalty for using small m in terms of efficiency is small compared to the advantages gained in simplicity and robustness. It should be noted that for $\beta_i = 0, T_i^*$ can be formed by a simple threshold process after the computation of H_N or no ranking of x_{ji} is required to calculate $F_{in_i} H_N^{-1}(a_j)$. Even if a small number of $\beta_i \neq 0$, the detector is almost as powerful, more robust and easier to implement than a comparable rank detector.

Corollary 1: The set $\{W_i(a_j)\}$, $j = 1,\ldots,m$ converges in law as $n_{min} \to \infty$ to a set of jointly normal random variables $\{W_{io}(a_j)\}$ where

$$E[W_{io}(a_j)] = F_i H^{-1}(a_j) \tag{77}$$

and

$$\mathrm{cov}[W_{io}(a_j), W_{io}(a_k)] = \sum_{\mu=1}^{M} n_\mu^{-1}(\delta_{i\mu} - \beta_\mu b_i(a_j))$$

$$\cdot(\delta_{i\mu} - \beta_\mu b_i(a_k) F_\mu H^{-1}(a_j)[1 - F_\mu H^{-1}(a_k)] \tag{78}$$

for $a_j \leq a_k$.

Proof: Following directly from Lemma 1 of the appendix, $n_{min}^{\frac{1}{2}} W_i(x)$ can be expressed as the sum of independent processes, each of which converges a.s. to a Wiener process. The corollary follows. The covariance between T_i^* and T_j^* follows from a similar calculation.

F. Efficiency of the Generalized Quantile Tests

The efficiency of the tests given by Equation (76) will be found in terms of the ARE with respect to the locally most powerful detector. Since the ARE is expressible as the ratio of two efficacies and the efficacy of the locally most powerful detector corresponds to Fisher's information in shift-of-mean problems (see Equation (14)), it remains to calculate the efficacy of the T_i^* tests.

It is recalled that the efficacy of a test T_i is

$$\mathcal{E}(T_i) = \frac{[\frac{\partial}{\partial\theta} E\{T_i\}\big|_{\theta=0}]^2}{\mathrm{var}(T_i)\big|_{\theta=0}} \tag{79}$$

where θ is a parameter defining the separation between F_1 and F_2 or θ is a relative measure of signal strength. For $F_1 = F_2 = \ldots = F_M$ the variance of Equation (75) reduces to

$$\operatorname{var}(T_i)\Big|_{\theta=0} = 2 \sum_{k=1}^{M} \frac{(\delta_{ik} - \beta_k)^2}{n_k} \int_0^1 \int_0^\alpha \beta(1 - \alpha)d\nu(\alpha)d\nu(\beta) \tag{80}$$

and for $M = 2$

$$\operatorname{var}(T_1)\Big|_{\theta=0} = A(1 - \beta_1)^2 \frac{(n_1 + n_2)}{n_1 n2} \tag{81}$$

where

$$A = 2 \int_0^1 \int_0^\alpha \beta(1 - \alpha)d\nu(\beta)d\nu(\alpha) \tag{82}$$

using assumption 1 of Theorem in this section, the order of differentiation and integration in Equation (74) may be interchanged or

$$\left[\frac{\partial}{\partial\theta} E\{T_i\}\Big|_{\theta=0}\right]^2 = \left[\int_0^1 \left[\frac{\partial}{\partial\theta} F_i M^{-1}(\alpha)\right]\Big|_{\theta=0} d\nu(\alpha)\right]^2 \tag{83}$$

The efficacy is then

$$E(T_i) = \frac{C^T BC}{C^T \Sigma_o C} \tag{84}$$

where $C^T = (C_1, \ldots, C_m)$ and B, Σ_o are $m \times m$ symmetric matrices with elements

$$b_{jk} = \frac{\partial W_i(a_j)}{\partial\theta}\Big|_{\theta=0} \cdot \frac{\partial W_i(a_k)}{\partial\theta}\Big|_{\theta=0} \tag{85}$$

$$\sigma_{jk} = \operatorname{cov}(W_i(a_j), W_i(a_k)) \tag{86}$$

respectively. By corollary 1, Equation (86) reduces for $\theta = 0$ to

$$\sigma_{jko} = \min(a_j), (a_k)[1 - \max(a_j, a_k)] \tag{87}$$

The scores which maximize the efficacy of Equation (84) may be calculated by forming a linear transformation

$$E(T_1) = Y_N^T(\Sigma^{-\frac{1}{2}}B\Sigma^{-\frac{1}{2}})Y_N \tag{88}$$

where

$$Y_N = \Sigma^{-\frac{1}{2}}C_N \ , \ Y^TY = 1$$

The efficacy corresponds to the largest eigenvalue of $B\Sigma^{-1}$ and the optimum scores are the components of the eigenvector associated with that eigenvalue. Further simplification of the efficacy expression is possible by noting that B is of rank 1 so that

$$E(T_1) = \text{tr}(\Sigma^{-\frac{1}{2}}B\Sigma^{-\frac{1}{2}}) \tag{89}$$

which reduces to the quadratic form

$$E(T_1) = F^T\Sigma^{-1}F \tag{90}$$

where $F^T = (f_1,f_2,\ldots,f_m)$ and

$$f_k = \frac{\partial}{\partial\theta} F_iH^{-1}(a_k) \Big|_{\theta=0} \tag{91}$$

where the notation $H_\theta^{-1}(\cdot)$ was introduced to emphasize the dependence of $H(\cdot)$ on the signal parameter θ.

For $M = 2$ (binary detection)

$$\beta_1F_1(H_\theta^{-1}(\lambda) + \beta_2F_2(H_\theta^{-1}(\lambda)) = \lambda$$

or

$$F_1H_\theta^{-1}(\lambda) = \lambda + \beta_2[F_1(H_\theta^{-1}(\lambda) - F_2(H_\theta^{-1}(\lambda))]$$

Taking the first term of the Taylor's series expansion

$$F_1H_\theta^{-1}(\lambda) \doteq \lambda + \theta\beta_2\{\frac{\partial}{\partial\theta} F_1H_\theta^{-1}(\lambda) - \frac{\partial}{\partial\theta} F_2H_\theta^{-1}(\lambda)\} \Big|_{\theta=0} \tag{92}$$

For the shift-of-mean alternatives

$$F_1(x) = F(x - \theta) \quad \text{and} \quad F_2(x) = F(x)$$

Equation (92) reduces to

$$\frac{\partial}{\partial\theta} F_1 H_\theta^{-1}(\lambda)\Big|_{\theta=0} \doteq \beta_2 f(F^{-1}(\lambda)) \tag{93}$$

For the Lehmann alternative which represents a useful class of stochastically ordered alternatives

$$F_1(x) = F^{1+\theta}(x) \quad \text{and} \quad F_2(x) = F(x)$$

Equation (92) reduces to

$$\frac{\partial}{\partial\theta} F_1(H_\theta^{-1}(\lambda))\Big|_{\theta=0} \doteq \beta_2 \lambda \ln\lambda \tag{94}$$

For the change-of-scale alternative

$$F_1(x) = F((1 + \theta)x) \quad \text{and} \quad F_2(x) = F(x)$$

Equation (92) reduces to

$$\frac{\partial}{\partial\theta} F_1(H_\theta^{-1}(\lambda))\Big|_{\theta=0} \doteq \beta_2 F^{-1}(\lambda) f(F^{-1}(\lambda)) \tag{95}$$

For the T_i^* statistics, the set of scores which maximizes the efficacy for a particular $\vec{a} = a_1, \ldots, a_m$ is easily found. A simple gradient search routine can be used to find the optimum $\vec{a}$ for a given $F(\cdot)$.

In Table 2, the efficacy of the T_1 statistic $(M = 2)$ is given for increasing values of m and optimum $\vec{a}$, where the parameter $(n_1 + n_2)/n_1 n_2$ has been suppressed. The last column is the maximum efficacy (Fisher's information for shift and Lehmann alternatives and, it can be shown, one plus the Fisher's information for change-of-scale alternatives). The efficacy of the best T_i^* test reaches 94% of the locally most powerful test for $m = 5$ and shift or Lehmann alternatives, and 84% for the Gaussian scale alternative with $m = 4$. Since the efficacy of the T_i^* tests for $M = 2$ is independent of the parameters β_1 and β_2, if we let $\beta_1 = 0$, it can be concluded that the two-sample quantile test has the same efficacy as the rank test while offering increased simplicity in implementation. Table 3 is included to demonstrate the robustness of the T_i^* tests. In the table, the test designed for one type of hypothesis testing problem has been applied to a different noise environment and the resulting loss in efficiency compared to the best T_i^* for that environment.

G. Concluding Remarks

In this section a unifying representation has been introduced which permits simultaneous analysis of linear rank and quantile tests. The generalization includes the study of weighted quantile tests which, in turn, permit the incorporation in the detector selection such design constraints as reliability of the available data, additional information about a particular type of data and other general information about the problem. Asymptotic normality of the class of detectors has been established including useful expressions for asymptotic mean and variance which lead to evaluation of efficacies.

Motivated by the desire for increased robustness and simplicity of implementation a particular subset, T_i^*, has been singled out for detailed study. The latter subset was found to be efficient as compared to locally most powerful detector, robust and relatively simple to implement, particularly if some β_i are zero.

VII. NEW DIRECTIONS

The area of partition tests which has been outlined in this chapter is in its infancy. There are numerous theoretical and practical problems which remain open. Among the problems that need immediate attention are inclusion of sample dependence in the analysis, evolution of simple and efficient learning and adaptive systems with application not only in the area of one-dimensional data processing (communications, radar, sonar, etc.) but also in the area of two and higher-dimensional data processing (image processing, pattern recognition, earth resources satellite data, metero-logical, oceanographic and biomedical data, etc.). In connection with that problem, additional techniques for estimation of quantiles and p.d.f.'s need be generated. In evolving the detection techniques, new objective and subjective measures of performance for small sample sizes need be introduced.

In addition, the representation theory and techniques of section VI are particularly promising as means of developing numerous new detectors which are simple, easy to implement, robust and powerful in specific applications by utilizing additional information about the problems under consideration.

Though the M-ary detection systems are discussed in the chapter, considerable additional research is needed in this area to assure effective utilization of the available bandwidth. It is hoped that the use of the new microprocessing techniques in conjunction with additional theoretical developments will result in a practical detector for future data transmission systems.

In conclusion, it is hoped that the material presented in this chapter will generate additional interest in the area of nonparametric tests based on quantile information. As was shown in the chapter, the performance of detectors based on this information is comparable to that of the rank detectors but is achievable by much simpler structures with real-time processing in most cases. In the adaptive sequential version of the quantile detectors, the additional sampling yields an updated test statistic without storage and re-ranking. Without any doubt, the study of detectors based on quantile and related information is well motivated and holds future promise.

APPENDIX: ASYMPTOTIC NORMALITY OF T_i

Let $\{x_j^i; j = 1,\ldots,n_i;\ i = 1,\ldots,M\}$ denote a vector of samples, n_i i.i.d. samples from F_i. F_{in_i} is the empirical distribution function of the n_i samples from F_i. Define

$$H_N(x) = \sum_{i=1}^{M} \beta_i F_{in_i}(x) \tag{96}$$

and

$$H(x) = \sum_{i=1}^{M} \beta_i F_i(x) \tag{97}$$

where

$$\sum_1^M \beta_i = 1\ ,\ \beta_i \geq 0\ ,\ \sum_1^M n_i = N$$

Define

$$T_i = \int_0^1 F_{in_i}(H_N^{-1}(\alpha))d\nu(\alpha) \tag{98}$$

In order to establish the asymptotic normality of (98), consider the process

$$L_{N,i}(t) = n_{min}^{\frac{1}{2}}\{F_{in_i}(H_N^{-1}(t)) - F_i(H^{-1}(t))\} \tag{99}$$

where

$$n_{min} = \min_{1\leq i\leq M}(n_i)$$

It will be shown that this process converges to a separable tied-down Wiener process in a manner which insures convergence of the integral of (99) with respect to ν to a zero mean random variable. Let D be the space of all right continuous functions having only jump discontinuities on [0,1]. Define the metrics

$$\rho(f,g) = \sup_{0\leq t\leq 1} |f(t) - g(t)| \qquad f,g \in D$$

and

$$\rho_q(f,g) = \sup_{0\leq t\leq 1} \left|\frac{f(t) - g(t)}{q(t)}\right| \qquad f,g \in D$$

where $q \in Q$ is defined in Section VI.

Let

$$u_i(t) = n_i^{\frac{1}{2}}[F_{in_i}(F_i^{-1}(t)) - t] \tag{100}$$

Let $u_o(t)$ be a separable tied-down Wiener process. Then, from Theorem 2.1 of [31],

$$\rho_q(u_i(t),u_o(t)) \underset{p}{\rightarrow} 0 \quad \text{as} \quad n_i \rightarrow \infty$$

where $u_o(t):0 \leq t \leq 1$ is a Gaussian process with

$$E\{u_o(t)\} = 0$$

and

$$E\{u_o(s)u_o(t)\} = s(1 - t) \qquad 0 \leq s \leq t \leq 1$$

Lemma 1. With probability one, $L_{N,i}$ can be expressed $\forall\ i:\beta_i \neq 1$ on (0,1] by

$$L_{N,i}(t) = \sum_{j=1}^{M} \left(\frac{n_{min}}{n_j}\right)^{\frac{1}{2}}(\delta_{ij} - \beta_j a_{N,i}(t))u_j(F_j H_N^{-1}(t)) + \delta_{N,i}(t) \tag{101}$$

where δ_{ij} is the Kronecker delta and

$$a_{N,i}(t) = \frac{[F_i H^{-1}(HH_N^{-1}(t)) - F_i H^{-1}(t))]}{[HH_N^{-1}(t) - t]} \tag{102}$$

$$\delta_{N,i}(t) = n_{min}^{\frac{1}{2}} a_{N,i}(t)[H_N H_N^{-1}(t) - t] \tag{103}$$

Proof: Using the same technique as Lemma 3.1 of [31], but changing the norming factor to $n_{min}^{\frac{1}{2}}$, (99) can be expanded

$$L_{N,i}(t) = \left(\frac{n_{min}}{n_i}\right)^{\frac{1}{2}} u_i(F_i H_N^{-1}(t)) + n_{min}^{\frac{1}{2}} a_{N,i}(t)[HH_N^{-1}(t) - t] \quad (104)$$

In (104) the second term has been divided by $[HH_N^{-1}(t) - t]$; this term is nonzero except at a finite number of points on (0,1). The last term can be expanded

$$a_{N,i}(t) \cdot n_{min}^{\frac{1}{2}}[HH_N^{-1}(t) - t]$$

$$= - \sum_{j=1}^{M} \beta_j \left(\frac{n_{min}}{n_j}\right)^{\frac{1}{2}} a_{N,i}(t) u_j(F_j H_N^{-1}(t)) + \delta_{N,i}(t) \quad (105)$$

Lemma 2. Define

$$m = \min_{\substack{1 \le i \le M \\ \{i:\beta_i \ne 0\}}} \{n_i\}$$

If

$$\lim_{n_{min} \to \infty} (n_{min}^{\frac{1}{2}}/n) = o(q^{-1}(1/m))$$

then

$$\lim_{n_{min} \to \infty} \rho_q(\overset{*}{\delta}_{N,i}, 0) \underset{a.s.}{\to} 0 \quad (106)$$

where $\overset{*}{\delta}_{N,i} = \delta_{N,i}$ on $[1/m, 1 - 1/m]$ and zero elsewhere.

Proof: By the left continuity of H_N^{-1},

$$|H_N H_N^{-1}(t) - t| \leqq 1/m$$

and

$$|\delta_{N,i}(t)| \le \beta_i^{-1} n_{min}^{\frac{1}{2}}/m$$

so

$$\lim_{n_{min}\to\infty} \rho_q(\delta^*_{N,i},0) \le \beta_i^{-1} \lim_{n_{min}\to\infty} (q^{-1}(1/m)n_{min}^{\frac{1}{2}}/m) = 0$$

Lemma 3. For every $\epsilon > 0$, there exists a $\beta > 0$ such that

$$\Pr\{F_i H_N^{-1}(t) \le \beta t \;\forall\; t \ge 1/m\} \ge 1 - \epsilon \qquad (107)$$

Proof: This is Lemma 2.5 of [31], but that proof required nonzero values of β_i . The following proof requires a new constraint, assumption 2 of Section VI. Define the event

$$A = \{F_i H_N^{-1}(t) \le \beta t \;\forall\; t \ge 1/m\}$$

If F_i is strictly increasing and we use the extended real line to define $F_i^{-1}(x) = \infty$ if $x \geqq 1$, this event is

$$\{H_N H_N^{-1}(t) \leqq H_N F_i^{-1}(\beta t) \quad 1/m \le t \leqq 1\}$$

but $[H_N H_N^{-1}(t) \le t + 1/m]$ so define the event $B \in A$

$$B = \{H_N F_i^{-1}(\beta t) \ge t + 1/m \qquad 1/m \le t \leqq 1\}$$

Now define the monotonic mapping from $[\beta/m, \beta]$ to $[x_1, \infty]$, where

$$x = F_i^{-1}(\beta t),\ u_\beta(x) = F_i(x)/\beta,\ x_1 = F_i^{-1}(\beta/m)$$

Define $C \in \beta$

$$C = \{F_{jn_j}(x) \ge \frac{u_\beta(x) + 1/m}{\beta_j} \;\forall\; j{:}\beta_j \ne 0,\ x_1 < x < \infty\}$$

but $F_{jn_j}(x)$ has the same distribution as $\Gamma_{n_j}(F_j(x))$ where Γ_{n_j} is the empirical distribution function from the sample of size n_j from the uniform distribution, so

$$C = \{\Gamma_{n_j}(t) \geq \frac{u_\beta(F_j^{-1}(t)) + 1/m}{\beta_j} \; ; \; F_j(x) < t \leq 1\}$$

or

$$C = \{\Gamma_{n_j}(t) \geq \frac{F_i F_j^{-1}(t)}{\beta\beta_j} + \frac{1}{m\beta_j} \; ; \; F_j F_i^{-1}(\beta/m) \leq t \leq 1\}$$

At this point, an assumption about the behavior of the F_i distributions is required. We need to assure that there exists a $C > 0$ such that

$$\frac{F_i F_j^{-1}(t)}{\beta\beta_j} + \frac{1}{m\beta_j} \leq ct \quad \text{on} \quad F_j F_i^{-1}(\beta/m) \leq t \leq 1$$

Assume that $F_i F_j^{-1}(t)$ has a finite slope d at $t = 0$ and there exists a constant f such that

$$F_i F_j^{-1}(t) \leq ft \quad \text{on} \quad 0 \leq t \leq 1$$

These two conditions insure the desired result. It is now possible to define $D \in C$

$$D = \{\Gamma_{n_j}(t) \geq ct \;\forall\; t:\Gamma_{n_j}(t) > 0\}$$

But it is well known that for every $\varepsilon > 0$ there exists a $c > 0$ such that

$$\Pr\{D\} \geq 1 - \varepsilon$$

Since $D \in A$, the lemma is complete.

Lemma 4. For $q \in Q$

$$\rho_q(u_i^*(F_i H_N^{-1}(t)) \; , \; u_o(F_i H^{-1}(t))) \underset{P}{\to} 0 \tag{108}$$

where U_i^* is equal to U_i on $[1/m, 1 - 1/m]$ and zero elsewhere.

Proof: As in Theorem 2.1 of [31], the problem reduces to a study of the supremum over intervals $[0,\delta]$ and $[1-\delta,1]$ for sufficiently small δ. For in the interior interval $(\delta, 1-\delta)$, $F_iH_N^{-1}(t)$ is bounded away from zero with high probability, and (101) can be applied in this interval.

Theorem 2.1 is followed exactly with the substitution of Lemma 2 of this appendix for Lemma 2.5 of [31] and by noting that

$$\Pr\{F_iH_N^{-1}(t) < \alpha \qquad 0 < t \leq \delta\} > 1 - \varepsilon$$

using the same arguments of Lemma 3 and assumption 2.

Lemma 5. Define

$$L_{0,1}(t) = \sum_{j=1}^{M} \alpha_j^{\frac{1}{2}}[\delta_{ij} - \beta_j a_i(t)]u_{0,i}(F_iH^{-1}(t)) \tag{109}$$

where

$$\alpha_j = \lim_{n_{min}\to\infty} \left(\frac{n_{min}}{n_j}\right), \quad a_i = \frac{\partial}{\partial t} F_iH^{-1}(t)$$

Define $L_{N,i}^*$ equal to $L_{N,i}$ on $[1/m, 1-1/m]$ and zero otherwise. Then

$$\rho_q(L_{0,i}, L_{N,i}^*) \underset{p}{\to} 0 \tag{110}$$

Proof: The proof follows exactly Section 4 of [31] with the substitution of Lemmas 1-3 for the corresponding Lemmas of [31]. Using the triangle inequality, each of the individual terms contributes a quantity bounded above by

$$\rho_q(L_{0,i}, L_{N,i}^*) \leq \sum_{j=1}^{M} \rho_{q,j}$$

where

$$\rho_{qj} = \sup_{0\leq t\leq 1} \Big| q^{-1}(t)\{\alpha_j^{\frac{1}{2}}[\delta_{ij} - \beta_j a_i(t)]u_{o,i}(F_iH^{-1}(t))$$

$$- \left(\frac{n_{min}}{n_j}\right)^{\frac{1}{2}}[\delta_{ij} - \beta_j a_{N,i}(t)]u_{N,i}^*(F_iH_N^{-1}(t))\}\Big|$$

But by assumption 1, $\rho(a_i, a_{N,i}) \underset{p}{\to} 0$, and by Lemma 3,

$$\rho_q(u_{o,i}(F_i H^{-1}), u^*_{N,i}(F_i H_N^{-1})) \underset{p}{\to} 0$$

so collecting terms and applying the product inequality repeatedly, the theorem is proved.

Theorem. Under conditions 1-5 in Section VI, let T_i' be the set of all T_i statistics for which $\beta_i \neq 1$, then $\{T_i'\}$ converge weakly as $n_{min} \to \infty$ to an $N(\overline{T}', \Sigma)$ random vector where

$$E\{T_i'\} = \int_0^1 F_i H^{-1}(\alpha) d\nu(\alpha) \tag{111}$$

$$\begin{aligned} \operatorname{cov}\{T_i' T_j'\} = 2n_{min}^{-1} \sum_{k=1}^{M} \alpha_k \int_0^1 \int_0^{\alpha} [\delta_{ki} - \beta_k a_i(t)] \\ \cdot [\delta_{jk} - \beta_k a_j(t)] F_k H^{-1}(\beta)[1 - F_k H^{-1}(\alpha)] d\nu(\beta) d\nu(\alpha) \end{aligned} \tag{112}$$

Proof: Define

$$T^*_{N,i} = \int_0^1 L^*_{N,i}(\alpha) d\nu_N(\alpha)$$

then

$$\begin{aligned} |T^*_{N,i} - \int_0^1 L_{o,i} d\nu| \leq |\int_0^{1/m} L_{o,i} d\nu| + \rho_q(\delta^*, 0) \int_0^1 q d|\nu| \\ + |\int_{1/m}^1 L_{o,i} d|\nu - \nu_N|| + \rho_q(L^*_{N,i}, L_{o,i}) \int_{1/N}^1 q d|\nu| \end{aligned}$$

The first term converges in probability to zero since $L_{o,i}$ is bounded above in probability. See Lemma 2.2 of [31]. The second term converges to zero by Lemma 2. The third is negligible by assumption and the fourth goes to zero by Lemma 5. So

$$\left|T^{*}_{N,i} - \int_0^1 L_{o,i}\right| \underset{p}{\rightarrow} 0$$

Since the variance of $\int L_{o,i} d\nu$ is finite, asymptotic normality is established by definition.

REFERENCES

1. Ching, Y.C. and L. Kurz (1972) "Nonparametric detection based on m-interval partitioning," IEEE Trans. on Inf. Th. IT-17: 251-257.
2. Gotz, B. and L. Kurz (1972) "Nonrank distribution-free detectors based on m-interval statistics," 1972 Internat'l Symp. on Inf. Th.
3. Gotz, B. (1971) "Quantile estimation, m-interval statistics and associated classification procedures," Ph.D. Dissertation, School of Engineering and Science, New York University.
4. Kersten, P. and L. Kurz, "M-interval detection of nonconstant signals," Progress Report No. 40, 15 Sept. 1974 through 14 Sept. 1975, Joint Services Technical Advisory Committee (JSTAC) Summary of Current Research, The Microwave Research Institute, Polytechnic Institute of New York.
5. Bushnell, W.J. and L. Kurz (1974) "The optimization and performance of detectors based on partition tests," 1974 Allerton Conf. on Circuits and Systems.
6. Bushnell, W.J. and L. Kurz, "Performance of linear quantile detectors in noise represented by a mixture distribution model," Progress Report No. 40, 15 Sept. 1974 through 14 Sept. 1975, JSTAC, Summary of Current Research, The Microwave Research Institute, Polytechnic Institute of New York.

7. Bushnell, W.J. and L. Kurz, "Selection of near-optimal partitions in m-interval detectors," Progress Report No. 40, 14 Sept. 1974 through 14 Sept. 1975, JSTAC, Summary of Current Research, The Microwave Research Institute, Polytechnic Institute of New York.

8. Bushnell, W.J. (1975) "The optimization and performance of detectors based on partition tests," Ph.D. Dissertation, Polytechnic Institute of New York.

9. Kurz, L. (1962) "A method of digital signalling in the presence of additive Gaussian and impulsive noise," 1962 I.R.E. Internat'l Conv. Record, Part 4: 161-173.

10. Cohen, S. and L. Kurz (1969) "An efficient nonparametric detection procedure using a computer selection technique," P.I.B. Symp. on Computer Processing in Communications XIX (M.R.I. Symposium Series): 617-628.

11. Kapp, J. and L. Kurz (1969) "Performance of two suboptimum detectors and signal selection in Gaussian and impulsive noise," Proc. Seventh Annual Allerton Conf. on Circuit and System Theory: 459-467.

12. Kapp, J. and L. Kurz (1970) "A class of detectors for a noise environment represented by the mixture distribution model," 1970 Internat'l Symp. on Inf. Th.

13. Brownstein, H. and L. Kurz (1970) "Parametric and nonparametric detectors based on a sequential sample median test," Inf. and Control 17(5): 417-435.

14. Silver, H. I. and L. Kurz (1970) "A class of continuous signal design problems in burst noise," IEEE Trans. on Inf. Th. IT-16(5): 570-581.

15. Silver, H. I. and L. Kurz (1972) "A class of discrete signal-design problems in burst noise," IEEE Trans. on Inf. Th. IT-18(2): 258-262.

16. Rappaport, S. S. and L. Kurz (1966) "An optimal nonlinear detector for digital data transmission through nongauggian channels," IEEE Trans. on Comm. Tech: 266-274.

17. Kersten, P. and L. Kurz, "Bivariable m-Interval Classifiers with application to Edge Detection", Information and Control, to appear.

18. Chadwick, H.D. and L. Kurz (1968) "Two sequential nonparametric detection procedures," Inf. and Control 13(5): 403-428.

19. Robbins, M. and S. Monro (1951) "A stochastic approximation method," Annals of Math. Stat. 22: 400-407.

20. Dvoretzky, A. (1956) "On stochastic approximation," Third Berkeley Symp. I: 39-55.

21. Sacks, J. (1958) "Asymptotic distribution of stochastic approximation," Annals of Math. Stat. 29: 373-405.

22. Burkholder, D.L. (1956) "On a class of stochastic approximation processes," Annals of Math. Stat. 27: 1049-1059.

23. Woinsky, M.N. and L. Kurz (1968) "An efficient nonparametric detector based on a three-sample classification model," IEEE Trans. on Inf. Th. IT-14: 744-752.

24. Woinsky, M.N. and L. Kurz (1969) "Sequential nonparametric two-way classification with a prescribed asymptotic error probabilities," Annals of Math. Stat. 40(2): 445-455.

25. Kurz, L. and G. Pasternack, "A quantile detectors in shift-of-mean and change-of-scale detection problems," Progress Report No. 40, 15 Sept. 1974 through 14 Sept. 1975, JSTAC, Summary of Current Research, The Microwave Research Institute, Polytechnic Institute of New York.

26. Pasternack, G. (1975) "Nonparametric classification using quadratic quantile statistics," Ph.D. Dissertation, Polytechnic Institute of

New York. Also 1976 Symposium on Information Theory (Kurz, L. and G. Pasternack).

27. Whittle, P. (1964) "On the convergence to normality of quadratic forms in independent variables," Teoriya Veroyatnosti i ee Primeneniya 9: 113-118.

28. Cochrane, J.I. and L. Kurz (1972) "Generalized sign-test classifiers within a unified framework of M-sample nonparametric tests," Proc. 1972 Princeton Conf. on Inf. Sci. and Systems: 479-483.

29. Cochrane, J.I. (1972) "Some robust procedures for classification and signal selection in unspecified noise environment," Ph.D. Dissertation, School of Engineering and Science, New York University.

30. Kurz, L. (1973) "Rank and nonrank nonparametric methodology in classification problems," 1973 Internat'l Symp. on Inf. Th.

31. Pyke, R. and G.R. Shorack (1968) "Weak convergence of a two-sample empirical process and a new approach to Chernoff-Savage theorems," Annals of Math. Stat. 39: 755-771.

32. Chernoff, M. and R.I. Savage (1958) "Asymptotic normality and efficiency of certain nonparametric test statistics," Annals of Math Stat. 29: 972-994.

33. Hajek, J. (1968) "Asymptotic normality of simple linear rank statistics under alternatives," Annals of Math. Stat. 39: 325-346.

34. Dupack, V. and J. Hajek (1969) "Asymptotic normality of simple linear rank statistics under alternatives, II," Annals of Math. Stat. 40: 1992-2017.

35. Govinparajulu, Z., L. LeCam, and M. Raghauachari (1965) "Generations of theorems of Chernoff and Savage on the asymptotic normality of test statistics," Fifth Berkeley Symp. on Math. Stat. and Prob. I: 609-638.

36. Doob, J.L. (1958) "Heuristic approach to the Kolnogorov-Smirnov theorems," Annals of Math. Stat. 29: 972-994.

37. Kersten, P. and L. Kurz,"Robustized Robbin-Monro algorithm with application to M-interval Detection", Information Sciences, October 1976 (to appear).

(a) Value of μ

	MODEL: Shift-of-Mean	MODEL: Change-of-Scale
distribution under H	$2(\delta\zeta)^2 \sum_{i=1}^{m} q_i^{-1}$	$2(\gamma\zeta)^2 \sum_{i=1}^{m} \Delta_i$
distributon under K	$-2(\delta\zeta)^2 \sum_{i=1}^{m} q_i^{-1}$	$2(\Psi\zeta)^2 \sum_{i=1}^{m} \nabla_i$

(b) Value of σ_T^2

	MODEL: Shift-of-Mean	MODEL: Change-of-Scale
distribution under H	$2\zeta^4 \mathrm{tr}(\underline{A}^2) + (2\zeta c)^2 \left[\frac{1}{3} \mathrm{tr}(\underline{A}^2) + 2\delta^2 \sum_{i=2}^{m-2} (q_i q_{i+1})^{-1}\right]$	$2\zeta^4 (\underline{A}^2) + (2\zeta c)^2 \left[\frac{1}{3} \mathrm{tr}(\underline{A}^2) + 2\gamma \sum_{i=1}^{m-3} \Delta_i \Delta_{i+1}\right]$
distribution under K	$2\zeta^4 \mathrm{tr}(\underline{B}^2) + (2\zeta c)^2 \left[\frac{1}{3} \mathrm{tr}(\underline{B}^2) + 2\delta^2 \sum_{i=2}^{m-2} (q_i q_{i+1})^{-1}\right]$	$2\zeta^4 \mathrm{tr}(\underline{B}^2) + (2\zeta c)^2 \left[\frac{1}{3} \mathrm{tr}(\underline{A}^2) + 2\Psi^2 \sum_{i=1}^{m} \nabla_i \nabla_{i+1}\right]$

TABLE 1

Alt. \ M	1	2	3	4	5	6	ε_{max}
Gaussian Shift	.64	.810	.883	.920	.942	.956	1.
Cauchy Shift	.405	.429	.434	.470	.479	.483	.5
Lehmann Alt.	.646	.820	.891	.927	.948	.961	1.
Gaussian Scale	.61	1.30	-	1.64	-	1.76	2.
Cauchy Scale	.143	.405	-	.461	-	.479	1.5

Table 2 Maximum efficacy of T_i^* statistics.

	Optimum Settings for Lehmann Alternatives		Optimum Settings for Gaussian Shift Alternative		Optimum Settings for Cauchy Shift Alternative	
M	Gaussian Noise	Cauchy Noise	Lehmann Alt.	Cauchy Noise	Lehmann Alt.	Gaussian Noise
1	1.11	5.65	1.29	0	1.29	0
2	1.04	4.74	1.15	2.55	1.89	0.62
3	1.00	4.45	1.08	2.74	3.36	3.20
4	1.06	4.76	1.02	3.24	4.47	2.99
5	0.96	4.89	0.99	3.36	4.47	2.99
6	0.93	4.96	0.97	3.44	4.47	3.01

Table 3 Loss in dB for 3 environments.

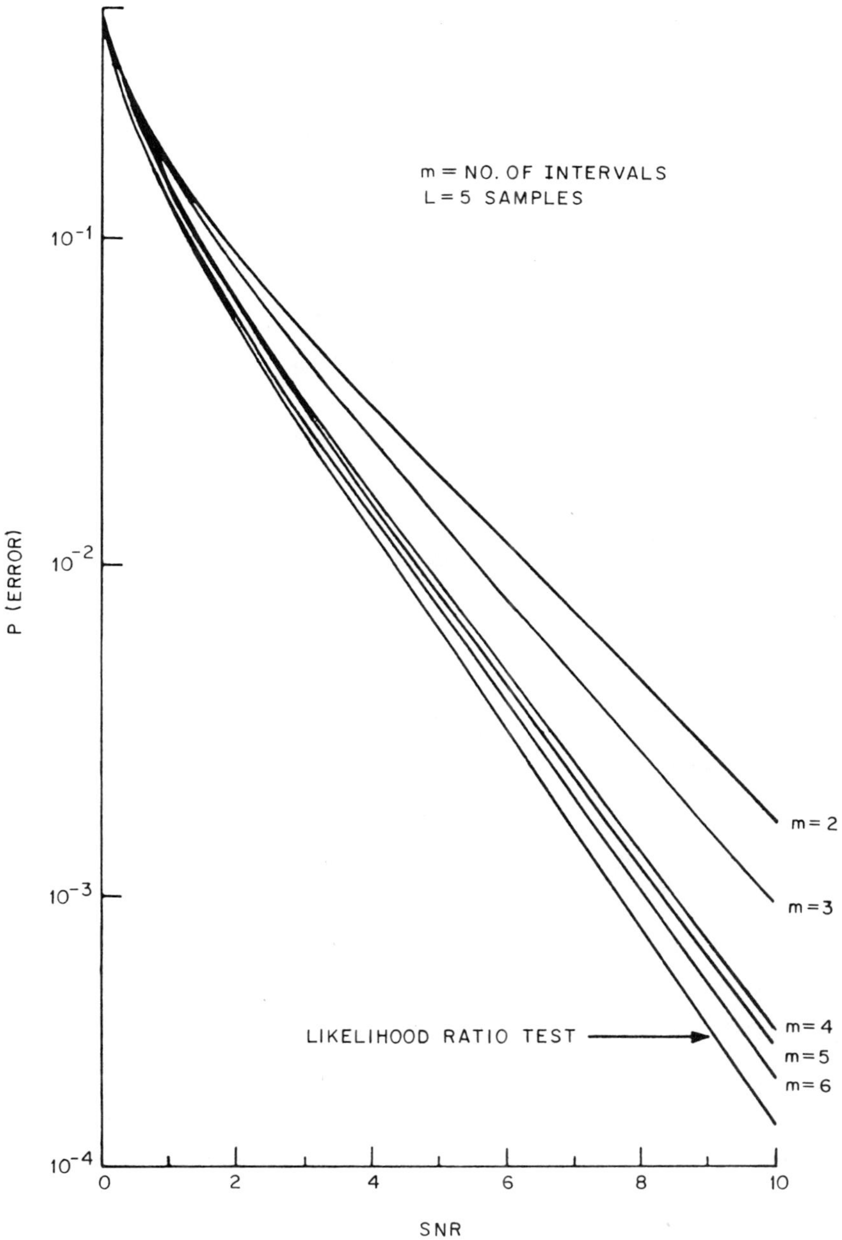

Figure 1 P{Error} vs SNR in Gaussian noise.

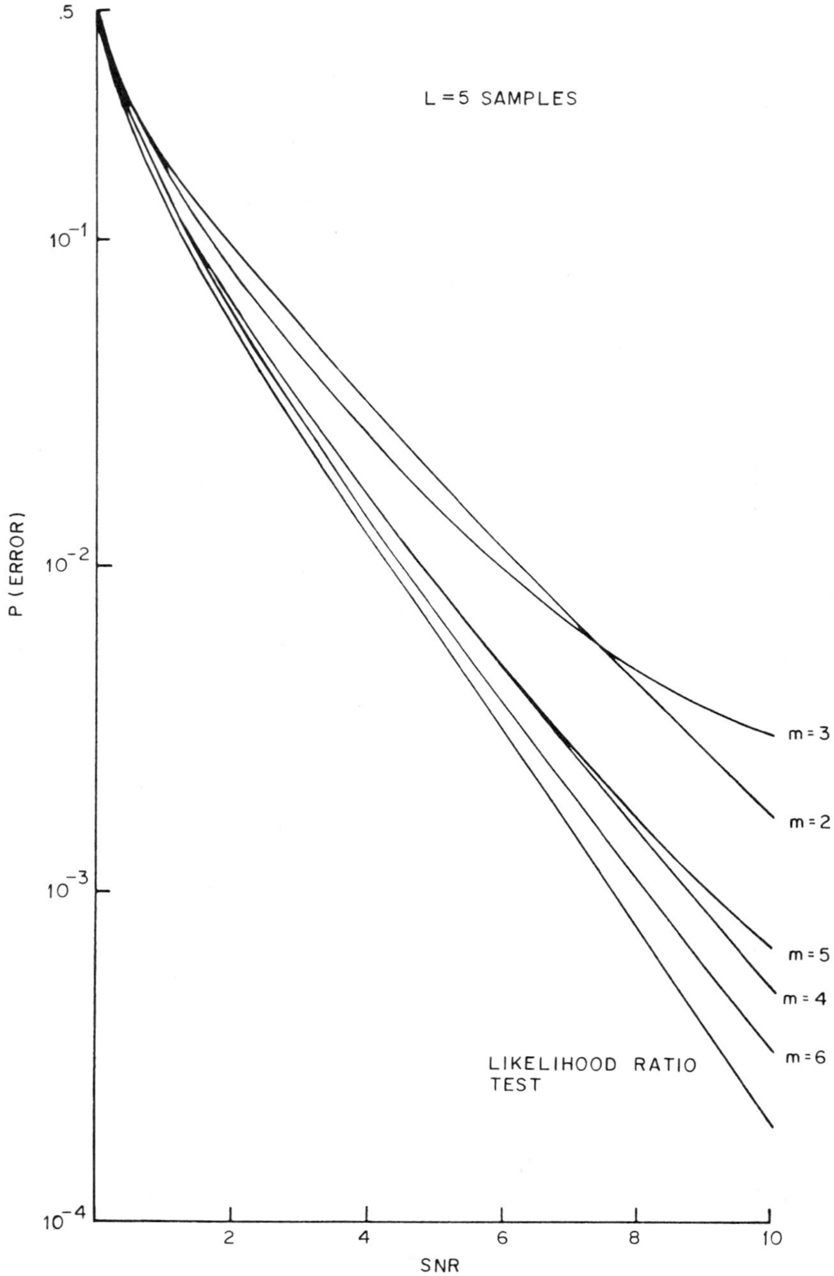

Figure 2 The P(error) of the partition test using the partitions which maximize the mutual information.

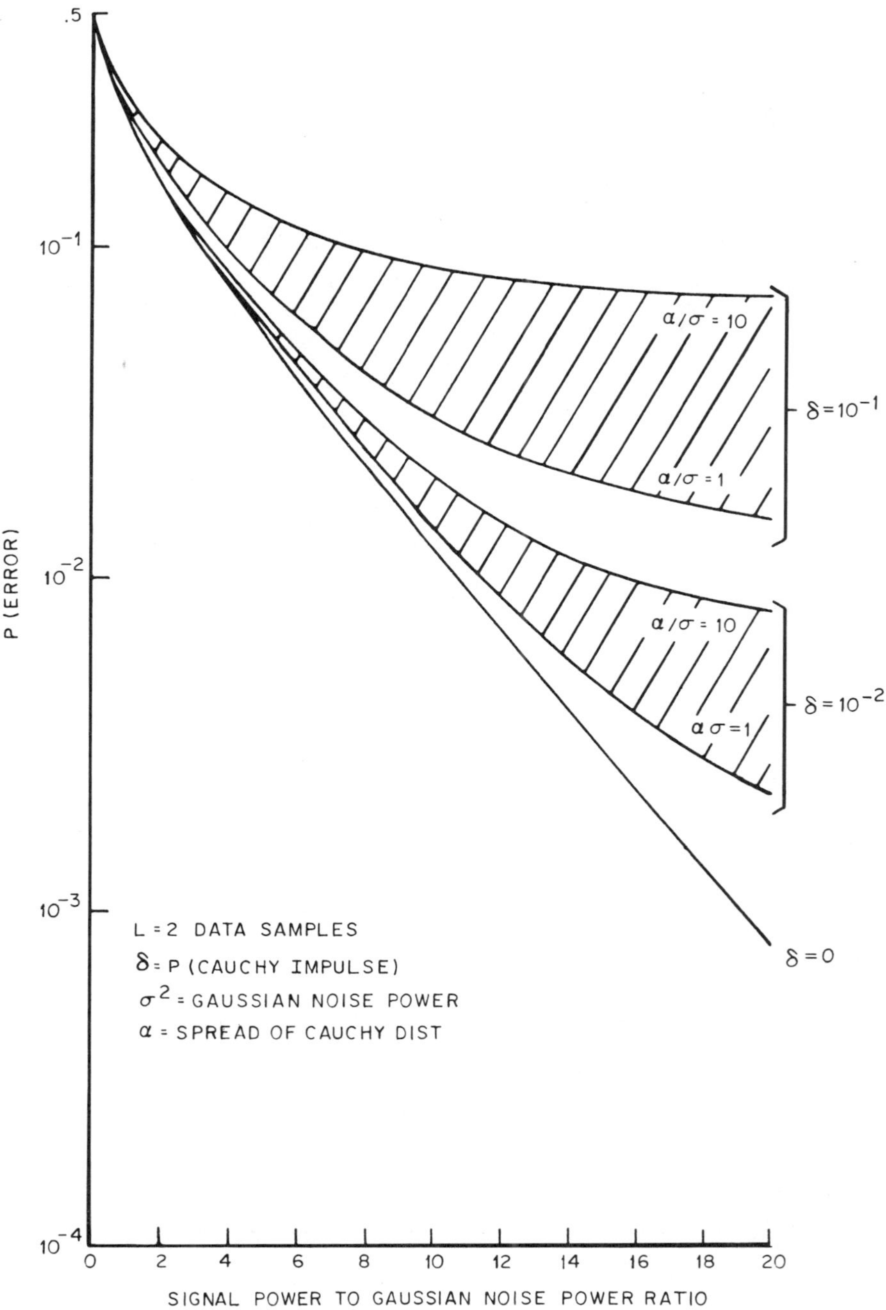

Figure 3 P{Error} of the Gaussian noise likelihood ratio test in a mixture of Gaussian and Cauchy noise.

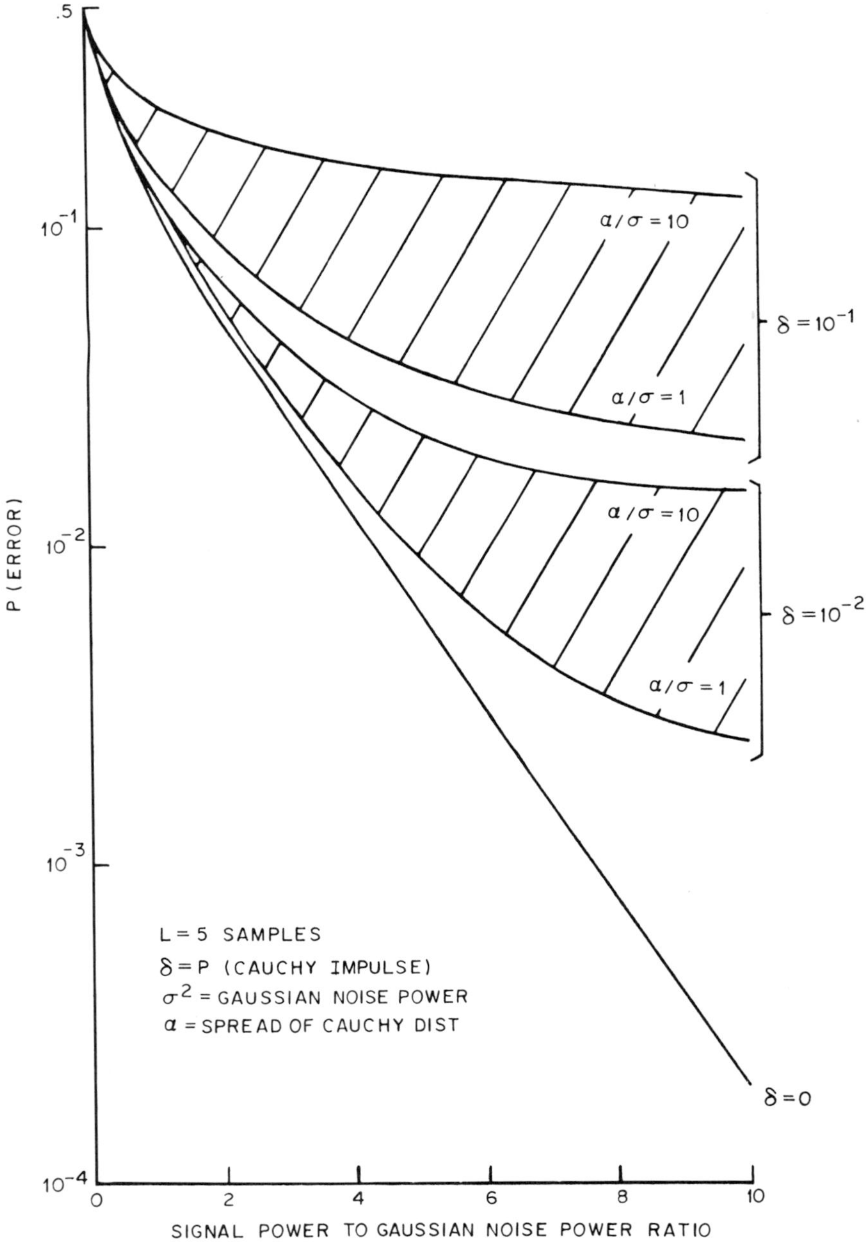

Figure 4 P{Error} of the Gaussian noise likelihood ratio test in a mixture of Gaussian and Cauchy noise.

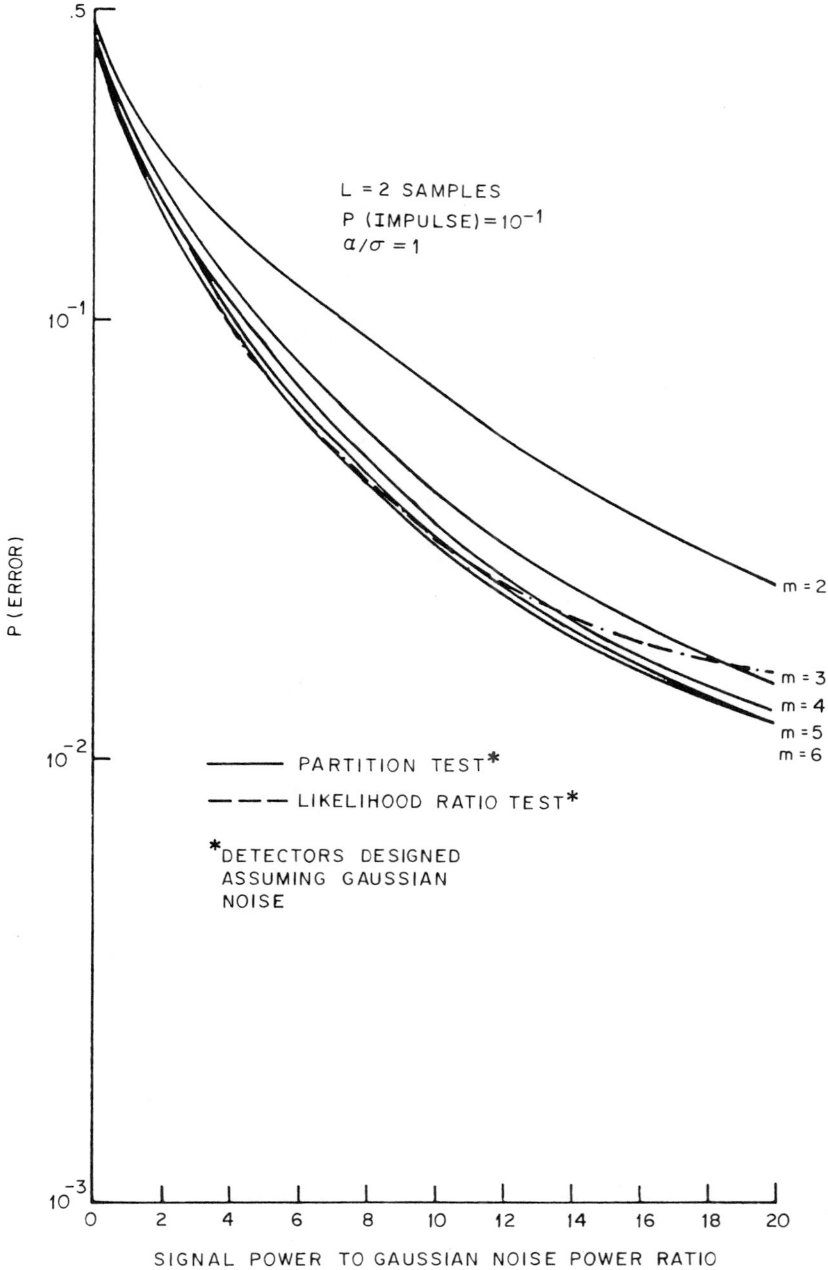

Figure 5 P{Error} of the Gaussian noise partition test in a mixture of Gaussian and Cauchy noise.

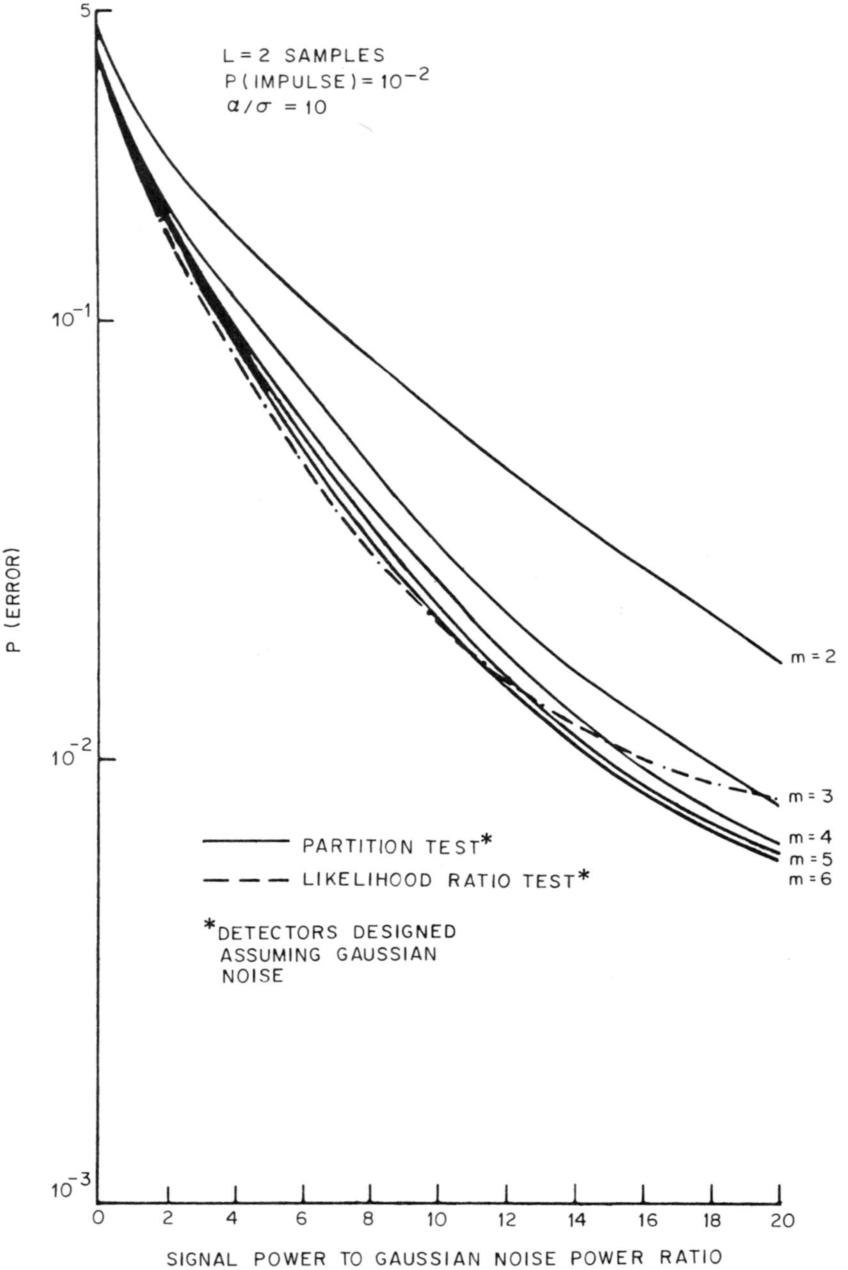

Figure 6 P{Error} of the Gaussian noise partition test in a mixture of Gaussian and Cauchy noise.

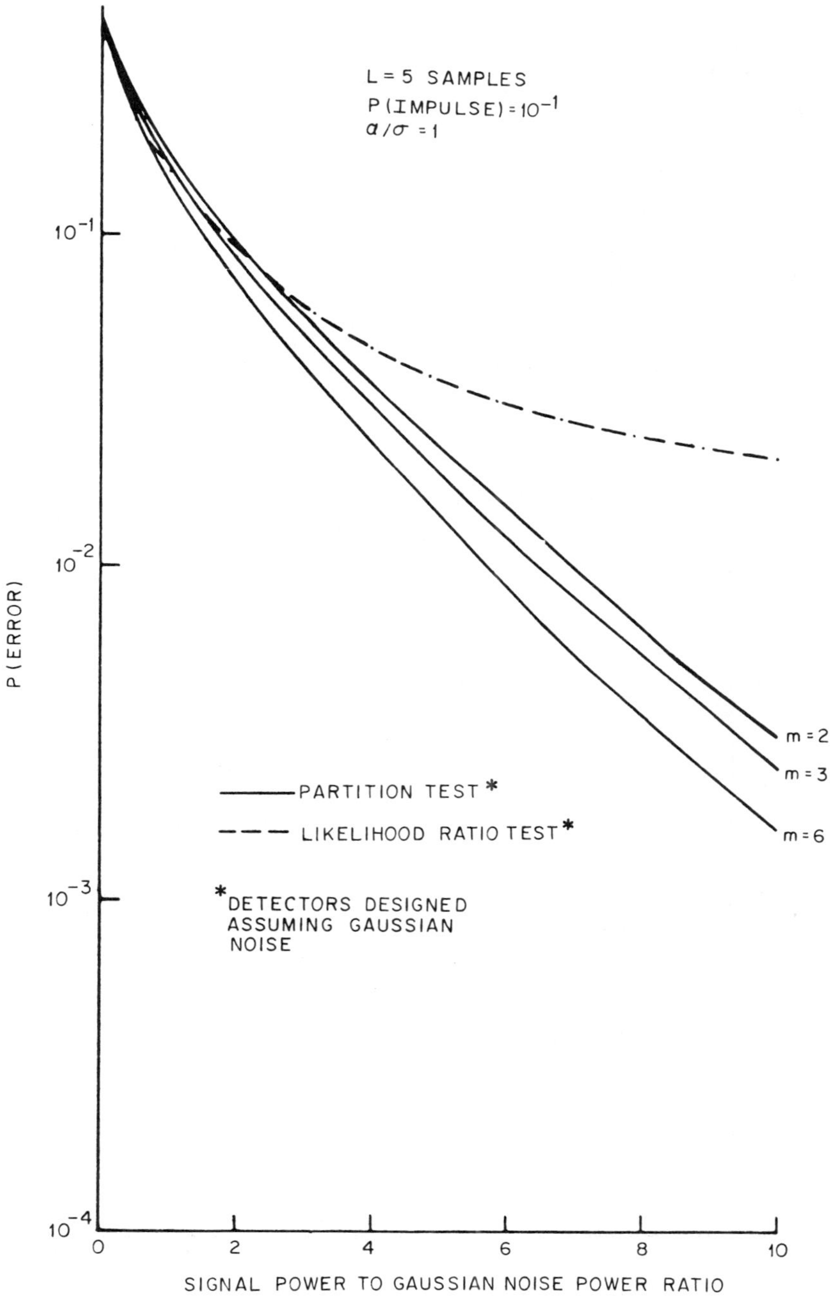

Figure 7 P{Error} of the Gaussian noise partition test in a mixture of Gaussian and Cauchy noise.

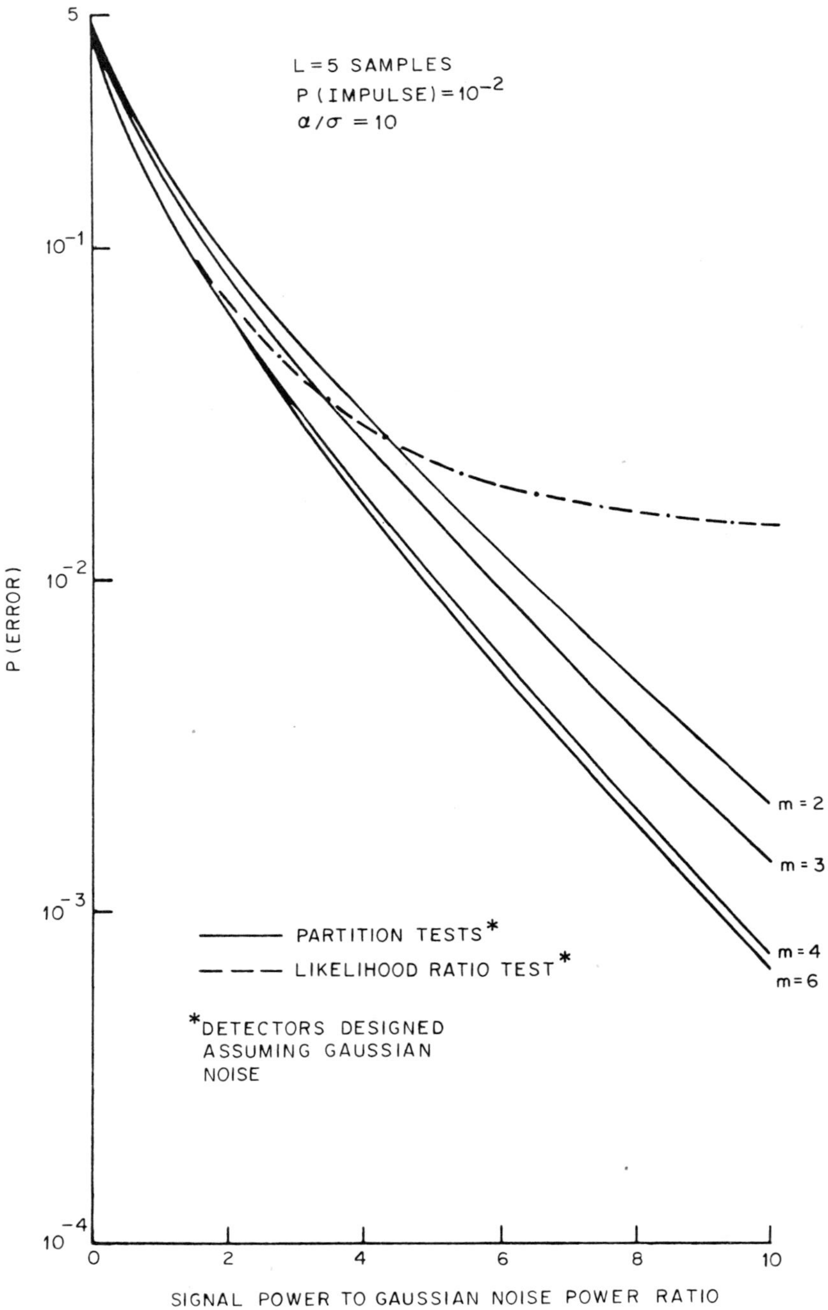

Figure 8 P{Error} of the Gaussian noise partition test in a mixture of Gaussian and Cauchy noise.

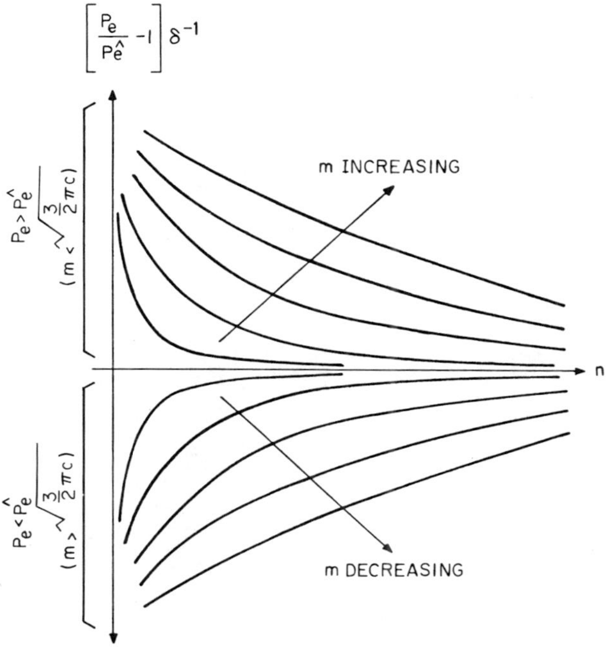

Figure 9 Performance ratio for fixed c = n/m .

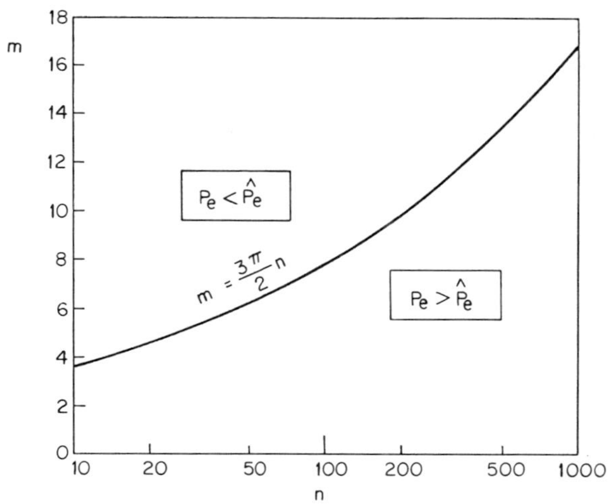

Figure 10 Equal performance boundary.

Chapter 4

CONDITIONAL TESTS IN NONPARAMETRIC DETECTION

Saleem A. Kassam

Department of Systems Engineering
Moore School of Electrical Engineering
University of Pennsylvania
Philadelphia, Pennsylvania

I. INTRODUCTION

From the time that a detailed analysis of some nonparametric statistical testing procedures for signal detection problems was first carried out by Capon [2], many different types of nonparametric signal detection methods have been studied. Reviews of the more common and useful of these techniques have been given by Thomas [28] and Carlyle [3]. In addition, several new methods have been proposed since these reviews were published [1,5,6,8,18].

In nonparametric detection theory it is commonly assumed that detector input data form discrete-time random processes, and the best-known tests are those based on signs (polarities) or ranks of the data. In general detectors such as the sign detector, which are easy to implement, suffer from low detection powers and efficiencies compared to the parametric detectors under assumptions of Gaussian or near-Gaussian input statistics. On the other hand, detectors based on rank tests have been shown to have very good performance characteristics for large classes of input statistics, but they are considerably more difficult to implement.

In this chapter we examine some new methods of nonparametric detection, which are based on the use of <u>conditional</u> statistical tests. Conditional testing is by no means a new concept in statistical theory; such testing procedures have been known to have a theoretical basis in the concepts of <u>sufficiency</u> and <u>completeness</u> of test statistics, and in the ideas of <u>similar</u> and <u>invariant</u> tests [9,23]. In a conditional test the threshold and randomization probability of a threshold test are not taken to be fixed parameters independent of the data, but are directly dependent on the specific data set being analyzed for a test of hypotheses. Thus the threshold and randomization probability are <u>conditioned</u> on the data. For a detailed treatment of the theoretical aspects of conditional testing, the reader is referred to the statistical literature [9,23]. In the following section a brief description of the major ideas of conditional testing is included, along with other statistical preliminaries.

In Sections III, IV, and V we discuss conditional-test detectors (that is, detectors based on conditional tests) which may be considered as being extensions and generalizations of the one-input sign detector, the two-input polarity coincidence correlator (PCC) and the one-input Wilcoxon detector, respectively. It will be seen that the use of conditional tests for construction of nonparametric signal detection systems leads to easily implemented detectors with very good performance characteristics. The performances are obtained both on the basis of asymptotic comparisons and from finite-sample results evaluated by exact computation of the detection probabilities.

In Section VI we discuss a specific application of conditional testing in the design of detectors for dependent input data. Finally, in Section VII, some directions for further research are indicated and other applications of conditional testing procedures are mentioned.

II. PRELIMINARIES

A. Conditional Tests

Consider a binary decision problem with a null hypothesis H and an alternative hypothesis K, and let $\underline{X}$ denote the n-component random observation vector. A fixed-threshold test for H against K is defined by a critical function $\Phi(\underline{X})$, through an appropriate test statistic $T(\underline{X})$, according to the equation

$$\Phi(\underline{X}) = \begin{cases} 1 & , T(\underline{X}) > t_{n,\alpha} \\ r_{n,\alpha} & , T(\underline{X}) = t_{n,\alpha} \\ 0 & , T(\underline{X}) < t_{n,\alpha} \end{cases} \tag{1}$$

The critical function $\Phi(\underline{X})$ is the probability with which the alternative K is accepted when $\underline{X}$ is observed. The threshold $t_{n,\alpha}$ and randomization probability $r_{n,\alpha}$ are chosen to obtain a desired probability of false alarm $P_{FA} = \alpha$, based on the distribution of $T(\underline{X})$ under H. The test statistic $T(\underline{X})$ is obtained from a consideration of the statistics of $\underline{X}$ under H and K, and is often the sum of a nonlinear function of the components of $\underline{X}$.

The test defined by equation (1) may be modified so that both the threshold and randomization probability become functions $t_{n,\alpha}(\underline{X})$ and $r_{n,\alpha}(\underline{X})$, respectively, of the observation vector. The resulting test is a conditional test, with values of the threshold test parameters being conditioned on the observations.

Conditional tests arise as a result of certain requirements which may be imposed on a testing procedure. For example, when H is a composite hypothesis a restriction to tests which are <u>similar</u> of size α [9] leads, by the Lehmann-Scheffé theorem [9,24], to a conditional test, when a <u>boundedly-complete</u> and <u>sufficient</u> statistic $S(\underline{X})$ exists under the probability measures of H. For a nonparametric hypothesis, the theory of <u>permutation tests</u> [13,27] can be applied to obtain a constant size α for all measures of H, when $\underline{X}$ (or a reduced data set) has a distribution invariant under certain groups of transformations on it. Such permutation tests usually result in conditional implementations.

As a simple example, consider the hypothesis of symmetry about the origin of $f(x)$, the density function of the components of $\underline{X} = (X_1, X_2, \ldots, X_n)$, assumed to be independent and identically distributed. Let $S(\underline{X})$ be the vector of magnitudes $(|X_1|, |X_2|, \ldots, |X_n|)$. Conditioned on $S(\underline{X})$, the observation vector can be any one of 2^n possibilities obtained by associating the 2^n permutations of positive and negative signs to the components of $S(\underline{X})$. Thus, a suitable test statistic can be defined for a particular alternative, and its conditional distribution, given $S(\underline{X})$, would be exactly known under H . A constant size-α threshold test may then be constructed, because

$$E_H\{\Phi(\underline{X})|S(\underline{X})\} = \alpha \tag{2}$$

implies

$$E_H\{\Phi(\underline{X})\} = \alpha \tag{3}$$

For the case of the normal, one-sided shift alternative, the conditional test discussed above with

$$T(\underline{X}) = \sum_{i=1}^{n} X_i$$

is in fact asymptotically power equivalent to the corresponding t-test from normal theory [9].

If the data is reduced to ranks and signs, once again the permutation argument holds (the $|X_i|$ are replaced by their relative ranks R_i). Now, however, the permutation test is not a conditional test. The vector of signs and the rank vector $(R_1,R_2,\ldots,R_n)$ are independent under the hypothesis of symmetry, and the signed-rank vector, for example, has a distribution under H which is exactly known.

The well-known rank tests may be viewed as being permutation tests not requiring a conditional implementation. Consideration of explicit conditional structures in constructing nonparametric detectors can, however, lead to useful and efficient extensions and generalizations of the common detectors, as we will see.

B. Asymptotic Performance Comparisons

To compare the performances of different detectors we will use both finite-sample, finite-signal computed results for the detection power β , as well as asymptotic performance comparisons based on their asymptotic relative efficiencies (AREs)[9,26]. The ARE of two detectors (or tests) is defined as the limiting ratio of sample sizes required by the two detectors to achieve the same performance in terms of power and false alarm probability, the limit being with respect to increasing sample sizes and vanishing signal strengths. In most cases, the ARE of two detectors can be obtained as a ratio of their <u>efficacies</u> [2,9]. The efficacy E of a test is defined as a limiting incremental signal-to-noise ratio,

$$E = \lim_{n\to\infty} \left[\frac{\frac{d}{d\theta} E_\theta\{T(\underline{X})\} \Big|_{\theta=0}}{n \operatorname{Var}_\theta\{T(\underline{X})\} \Big|_{\theta=0}} \right]^2 \tag{4}$$

where θ is a parameter (signal) giving the "distance" of the hypothesis H from K. The ARE of two detectors gives a useful indication of relative performance in cases where exact finite-sample results are difficult to obtain.

Since the detectors we consider here are based on conditional tests, the use of efficacies and AREs for comparisons has to be justified. Some theorems by Hoeffding [15] give the assumptions under which the power of a conditional test approaches that of a corresponding test not having a conditional threshold. Fraser [9, Chapter 7] gives an exposition of these concepts. The basic idea is that asymptotic comparisons can be made using an asymptotically equivalent parametric detector not using conditional testing. In our applications we will use these results to obtain AREs from the efficacies of detector test statistics, even when the threshold test is a conditional test.

C. Statistical Hypotheses for the Signal Detection Problems

We will be concerned with two signal detection problems: the detection of a known signal in additive noise, a one-input situation, and the detection of a common random signal additive on independent noises in two channels. Except in Section VI, observation vectors will be assumed to be vectors of components which are independent under null hypotheses of the absence of signals. Thus if observations are obtained as samples from continuous-time processes, the sampling rate is constrained to be low enough to result in independent samples. In Section VI we will examine a situation where this is not assumed. We also make the additional assumption that univariate density functions of the noise exist, and are symmetric about the origin. The mild symmetry assumption is the only specific assumption we make about the noise densities. For analysis, it is also assumed that the components (samples) forming our observation vector are identically distributed under the null hypotheses.

For the one-input case, the null hypothesis H_1 about the density function $p(\underline{x})$ of the observation vector $\underline{X}$ is

$$H_1 : p(\underline{x}) = \prod_{i=1}^{n} f(x_i) \ , \ f(x) = f(-x) \tag{5}$$

where $\underline{x} = (x_1, x_2, \ldots, x_n)$. In the two-input case where the observation vectors are $\underline{X} = (X_1, X_2, \ldots, X_n)$ and $\underline{Y} = (Y_1, Y_2, \ldots, Y_n)$, the null hypothesis H_2 for the joint density function of $\underline{X}$ and $\underline{Y}$ is

$$H_2 : p(\underline{x}, \underline{y}) = \prod_{i=1}^{n} f(x_i) f(y_i) \ , \ f(x) = f(-x) \tag{6}$$

In the one-input case, the alternative hypothesis K_1 describing signal presence is

$$K_1 : p(\underline{x}) = \prod_{i=1}^{n} f(x_i - \theta s_i) \ , \ f(x) = f(-x) \tag{7}$$

where the s_i are known signal components and the amplitude θ has any positive value. Corresponding to H_2, we have in the two-input situation the alternative K_2 given by

$$K_2 : p(\underline{x},\underline{y}) = \prod_{i=1}^{n} \int_{-\infty}^{\infty} f(x_i - s) f(y_i - s) dF_S(s) \tag{8}$$

where $F_S(s)$ is a distribution function describing the common random signal component.

For the pair (H_1, K_1), the detectors we discuss are based on the further assumption that the s_i have the same positive value. The tests can be extended in a trivial way to cover the general case.

D. A Class of Non-Gaussian Densities

In addition to comparing the performances of different nonparametric detectors under conditions of noise with Gaussian densities, comparisons under cases of non-Gaussian noise have to be made to establish the general superiority, or otherwise, of different detectors. A useful class of non-Gaussian densities which can be used for such comparisons is the class of generalized Gaussian densities [16,17,25]. Densities $f_k(\cdot)$ in this class are parameterized by a single parameter k and are defined by

$$f_k(x) = \frac{k}{2\Gamma(1/k)A(k)} \exp\{-[|x|/A(k)]^k\} \text{ , } k > 0 \tag{9a}$$

where

$$A(k) = [\sigma^2 \Gamma(1/k)/\Gamma(3/k)]^{\frac{1}{2}} \tag{9b}$$

In equation (9), σ^2 is the variance of $f_k(\cdot)$ and $\Gamma(\cdot)$ is the gamma function. This class of densities includes the Gaussian density, obtained with $k = 2$, and the double-exponential density, obtained with $k = 1$. In general, k corresponds to the rate of exponential decay of the tails of the densities. This class thus models a wide range of noise characteristics in addition to being analytically simple to handle. Use will be made of this class of densities to provide performance results for our detectors under non-Gaussian noise conditions.

III. GENERALIZATIONS OF THE SIGN DETECTOR

Consider the problem of testing H_1 versus K_1. The sign detector [28] can be used as a nonparametric detector for this problem. This detector is based on the test statistic $T_0(\underline{X})$ given by

$$T_0(\underline{X}) = \sum_{i=1}^{n} \ell_0(X_i) \tag{10}$$

where

$$\ell_0(x) = \begin{cases} 1, & x \geq 0 \\ -1, & x < 0 \end{cases} \tag{11}$$

It is well-known that the performance of the sign detector is quite poor relative to the performance of the linear detector when the density function $f(\cdot)$ is a Gaussian density function. The explanation for this is that the function $\ell_0(\cdot)$ is a poor (2-level) approximation to the linear function $\ell(x)$, defined as

$$\ell(x) = x ,$$

on which the linear detector test statistic is based. Thus better performance can be expected, at least for Gaussian noise densities and noise densities near the Gaussian, if the nonlinearity $\ell_0(\cdot)$ is replaced by one giving a closer approximation to the function $\ell(x)$.

This is the reasoning which leads us to consider the detectors in this section. We obtain generalizations of the sign detector which are based on higher-level quantizations of the inputs, resulting in more powerful detectors. To maintain nonparametric performance, that is, a constant false alarm probability for all measures of H_1, conditional testing is employed. The resulting detectors remain fairly simple to implement, so that they are much more attractive than the simple sign detector in many applications.

Generalizations of detectors based on polarities have been considered previously [5,12,22]; however, their nonparametric performance cannot be established without further specific assumptions about the noise distribution functions. For example, in [5], knowledge of a few quantiles of the noise distribution is assumed to be available, or has to be acquired through a learning process. Grouped rank tests for the hypothesis of symmetry have also been considered in the statistical literature [10,11]. These grouped tests are very similar to what we call generalized sign tests, again with the exception that conditional testing was not used to construct nonparametric grouped rank tests.

We first consider a very simple extension of the sign detector, the dead-zone limiter detector, and then consider a general formulation of extensions of the sign detector to (2k)-level detectors.

A. A Simple Three-Level Detector

A simple extension of the sign detector has been analyzed in [19] and illustrates clearly the advantages and ease of implementation of detectors based on conditional testing. This three-level detector, the dead-zone limiter detector, reduces input data through a three-level quantization instead of the binary quantization employed in the sign detector. The dead-zone limiter detector uses only the signs of samples X_i which satisfy $|X_i| > c$ for some fixed $c \geq 0$. This detector may be considered to employ the dead-zone limiter characteristic $\ell_c(\cdot)$ to operate on the data, where

$$\ell_c(x) = \begin{cases} 1, & x \geq c \\ 0, & -c \leq x < c \\ -1, & x < -c \end{cases} \tag{12}$$

The dead-zone limiter $\ell_c(x)$ is an extension of the hard limiter $\ell_0(\cdot)$

of equation (11) and reduces to it when $c = 0$. The test statistic $T_c(\underline{X})$ of the dead-zone limiter detector is given by

$$T_c(\underline{X}) = \sum_{i=1}^{n} \ell_c(X_i) \tag{13}$$

and may also be written in the form

$$T_c(\underline{X}) = 2N_{cp} - N_c \tag{14}$$

if the random variables N_c, N_{cp} are defined by

$$N_c = \sum_{i=1}^{n} |\ell_c(X_i)| \tag{15}$$

$$N_{cp} = \tfrac{1}{2} \sum_{i=1}^{n} [\ell_c(X_i) + |\ell_c(X_i)|] \tag{16}$$

From equations (15) and (16), N_c is the total number of samples outside $[-c,c)$ and N_{cp} is the total number of samples in $[c,\infty)$.

The sign detector has a constant false alarm probability; that is, it is _similar_ to size α for the hypothesis H_1. The distribution function of its test statistic under H_1 is completely specified, being obtained from the binomial distribution with parameters ½ ("probability of success") and n ("number of trials"). In the dead-zone limiter detector, both N_c and N_{cp} are random variables, being functions of $\underline{X}$, whereas in the sign detector $(c = 0)$ the number N_{0p} is always equal to n. A critical function based only on the statistic $T_c(\underline{X})$ cannot therefore be the basis for a nonparametric detector for H_1 versus K_1; the distribution of $T_c(\underline{X})$ under H_1 does depend on the density $f(\cdot)$. However, by going to a conditional test, a simple nonparametric detector can be constructed. Although the distribution of N_{cp} under H_1 depends on the specific $f(\cdot)$, the distribution of N_{cp} _conditioned_ on N_c is again binomial, with parameters ½ and N_c. Therefore a critical function [equation (1)] for which the functions $t_{n,\alpha}(\underline{X})$ and $r_{n,\alpha}(\underline{X})$ depend on the observations through N_c can be made the basis for a nonparametric detector. Such a critical function can be described by

$$\Phi(\underline{X}) = \begin{cases} 1 & , N_{cp} > t_\alpha(N_c) \\ r_\alpha(N_c) & , N_{cp} = t_\alpha(N_c) \\ 0 & , N_{cp} < t_\alpha(N_c) \end{cases} \qquad (17)$$

The values of $r_\alpha(n_c)$ and the thresholds $t_\alpha(n_c)$ are such that for each realization n_c of the random variable N_c, we have $E_{H_1}\{\Phi(\underline{X})|N_c\} = \alpha$. This implies that $E_{H_1}\{\Phi(\underline{X})\} = \alpha$ and thus a constant false alarm probability can be obtained.

In implementing the dead-zone limiter detector a table of values of $r_\alpha(n_c)$ and $t_\alpha(n_c)$ for $0 \leq n_c \leq n$ is required. This table can be generated and stored easily. It should be noted that the values of $r_\alpha(n_c)$ and $t_\alpha(n_c)$ do not depend on the particular value of c, which may be chosen to obtain good performance for a specific class of density functions. The values of $r_\alpha(n_c)$ and $t_\alpha(n_c)$ are also independent of the sample size n; therefore they may be considered to be fixed parameters of the detector.

Thus the dead-zone limiter detector remains nonparametric for any $c \geq 0$. Figure 1 shows the structure of the dead-zone limiter detector. In Figure 2, the sign detector structure is shown for comparison. The table look-up adds to the complexity of the dead-zone limiter; however, as we now indicate, the performance of this detector may be decidedly superior to that of the sign detector. Furthermore, the table look-up is not difficult to implement in a practical system.

For this simple conditional-test detector, the efficacy (under the alternative of a constant signal) can be obtained fairly easily. Since the test statistic $T_c(\underline{X})$ is given by equation (14), its conditional mean and variance, given N_c, can be obtained from $E_\theta\{N_{cp}|N_c\}$ and $\mathrm{Var}_\theta\{N_{cp}|N_c\}$. It is easy to see that with $F(\cdot)$ the distribution corresponding to $f(\cdot)$,

$$E_\theta\{N_{cp}|N_c\} = \left[\frac{1 - F(c - \theta)}{1 - F(c - \theta) + F(-c - \theta)}\right] N_c \tag{18}$$

and

$$\mathrm{Var}_\theta\{N_{cp}|N_c\} = \left(\frac{[1 - F(c - \theta)]F(-c - \theta)}{[1 - F(c - \theta) + F(-c - \theta)]^2}\right) N_c \tag{19}$$

Assuming that $F(c) < 1$, the strong law of large numbers implies that $N_c \to \infty$ and

$$\frac{N_c}{n} \to 1 - F(c - \theta) + F(-c - \theta)$$

with probability one, as $n \to \infty$. We can therefore conclude that the conditional distribution of $T_c(\underline{X})$, suitably normalized, is asymptotically standard normal. Using these results in equation (4), we obtain the efficacy E_{DZ} of the dead-zone limiter detector as

$$E_{DZ} = \frac{2f^2(c)}{1 - F(c)} \tag{20}$$

(with an additional assumption that $f(c)$ is continuous at c).

For the sign detector, the efficacy E_S is given by

$$E_S = 4f^2(0) \tag{21}$$

This also follows from equation (20) with $c = 0$. Finally, for the linear detector, which uses as a test statistic the sum of the n observations, the efficacy E_L is

$$E_L = 1/\sigma^2 \tag{22}$$

where σ^2 is the variance of the noise density $f(\cdot)$. The linear detector is not a nonparametric detector, but it is the locally optimum detector for H_1 vs K_1 when $f(\cdot)$ is a Gaussian density function.

We can now obtain $ARE_{DZ,L}$ and $ARE_{DZ,S}$, the AREs of the dead-zone limiter detector relative to the linear detector and sign detector, respectively, by forming the ratios E_{DZ}/E_L and E_{DZ}/E_S. We have already remarked that the dead-zone limiter detector remains nonparametric for any choice of $c \geq 0$; therefore, its asymptotic performance can be optimized for particular $f(\cdot)$ by choosing c optimally. For sufficiently smooth $f(\cdot)$ the equations for the optimum c can be obtained easily and can be readily solved numerically. Figure 3 shows $ARE^*_{DZ,L}$ and $ARE^*_{DZ,S}$ plotted against k, which parameterizes the class of generalized Gaussian densities of equation (9). The $*$ denotes optimum values obtained by using the maximizing values of c for each k. Also plotted in Figure 3 are the $ARE_{DZ,L}$ and $ARE_{DZ,S}$ for the fixed value of c which is the optimum for the case of Gaussian noise. (This value is 0.612σ). For comparison, $ARE_{S,L}$ is also shown. It can be seen that, for Gaussian noise, the dead-zone limiter detector is significantly superior to the sign detector (in terms of asymptotic performance). Furthermore, significant improvements in performance are obtained for densities close to the Gaussian in the class of non-Gaussian densities defined by equation (9), even without using optimum values of c.

An interesting result on the worst-case asymptotic performance of the dead-zone limiter detector relative to the parametric linear detector can also be established. Consider $ARE_{DZ,L}$, the ARE of the dead-zone limiter detector relative to the linear detector:

$$ARE_{DZ,L} = \frac{2\sigma^2 f^2(c)}{1 - F(c)} \tag{23}$$

We can look for the density function which minimizes this ARE optimized with respect to c; that is, we search for the least favorable symmetric density $f_L(\cdot)$, with associated distribution $F_L(\cdot)$, such that

$$\max_{c\geq 0} \frac{2f_L^2(c)}{1 - F_L(c)} \leq \max_{c\geq 0} \frac{2f^2(c)}{1 - F(c)} , \quad f(\cdot) \in \mathcal{F}_1 \tag{24}$$

where $\mathcal{F}_1$ is the class of all symmetric densities with unit variance and absolutely continuous distributions. In equation (24), we assume $\sigma^2 = 1$ without loss of generality. It is not difficult to show [the proof uses the lemma on p. 150 of reference 7] that the triangular density function (with unit variance), given by

$$f_L(x) = \begin{cases} 1/\sqrt{6} - |x|/6 , & |x| \leq \sqrt{6} \\ 0 , & |x| > \sqrt{6} \end{cases} \tag{25}$$

satisfies equation (24), and that the resulting worst-case (minimax) ARE in equation (23) is 2/3, achieved for any c in $[0,\sqrt{6})$. This can be compared with the corresponding value of 1/3 for the sign detector and the class of unimodal and symmetric densities [14]. This result implies that the use of a dead-zone limiter detector instead of the linear detector in the case of known $f(x)$ results in a maximum loss of efficiency of 1/3. Of course, in many cases an improvement in performance may be obtained. The advantage of the dead-zone limiter detector is that it is nonparametric and the thresholds for any false alarm probability do not depend on the particular $f(\cdot)$, even for finite sample sizes.

The finite-sample-size performance characteristics of the dead-zone limiter detector can be obtained by computing the exact detection probabilities for different signal strengths and sample sizes, with tables of $r_\alpha(n_c)$ and $t_\alpha(n_c)$ being used to achieve desired false alarm probabilities. Since the binomial probability distribution function is involved in such computation and table generation, a digital computer can be used for the purpose. Figures 4 and 5 show the results obtained for some typical false alarm probabilities. In the figures, detection probability β is plotted against normalized signal strength θ/σ for different sample sizes for the linear, sign, and dead-zone limiter detectors. The noise is assumed to be Gaussian and, in the latter detector, the value of c maximizing its asymptotic performance is used. From these results it is clear that, for Gaussian noise, the dead-zone limiter detector has a definitely better performance than the sign detector for finite samples and signals, as is true in the asymptotic case.

In Figure 6, detection probability for the dead-zone limiter detector is plotted against c/σ for fixed values of sample size, false alarm probability, and signal strengths. The flatness of these curves shows the relative insensitivity of the performance for deviations of c from the optimum value and indicate that the selection of c is not very critical.

Our discussion of the dead-zone limiter detector has brought out the main features of conditional testing. We now look at a more general formulation of such extensions of the sign test.

B. The (2k)-Level Extension of the Sign Test

In [20] and [21] further generalizations of the sign detector, based on conditional testing, have been described and we now consider these detectors. The general formulation we will discuss includes the dead-zone limiter detector as a particularly simple special case.

Consider the symmetric (2k)-level quantizer $\ell_{\underline{c},\underline{h}}(\cdot)$ of Figure 7. The vectors $\underline{c} = (c_1, c_2, \ldots, c_k)$ and $\underline{h} = (h_1, h_2, \ldots, h_k)$ are k-dimensional vectors characterizing this nonlinearity and according to Figure 7 we always set $c_k = 0$ and $h_1 = 1$, without loss of generality. In addition, c_0 is defined to be $+\infty$. The (2k)-level sign detector is based on a test statistic $T_{\underline{c},\underline{h}}(\cdot)$ given by

$$T_{\underline{c},\underline{h}}(\underline{X}) = \sum_{i=1}^{n} \ell_{\underline{c},\underline{h}}(X_i) \tag{26}$$

This reduces to $T_c(\underline{X})$ of equation (13) if we consider the case $k = 2$ with $c_1 = c$ and $h_2 = 0$. The specific choice $h_2 = 0$ results in a three-level quantization rather than the four-level quantization which is generally obtained with $k = 2$.

By defining new random variables N_{in}, N_{ip}, N_i for each integer i between 1 and k, we may rewrite equation (26) as

$$T_{\underline{c},\underline{h}}(\underline{X}) = \sum_{i=1}^{k} h_i(N_{ip} - N_{in}) = \sum_{i=1}^{k} h_i(2N_{ip} - N_i) \tag{27}$$

with

N_{ip} = the number of components X_i of $\underline{X}$ in $[c_i, c_{i-1})$ (28a)

N_{in} = the number of components X_i of $\underline{X}$ in $[-c_{i-1}, -c_i)$ (28b)

and

$$N_i = N_{in} + N_{ip} \tag{28c}$$

Thus N_{ip}, N_{in} are the numbers of samples falling in the ith positive and negative input intervals, respectively.

Conditioned on the vector $\underline{N} = (N_1, N_2, \ldots, N_k)$, the distribution of $T_{\underline{c},\underline{h}}(\underline{X})$ is completely specified under H_1, the hypothesis of symmetry. Thus a nonparametric detector for H_1 versus K_1 can be based on $T_{\underline{c},\underline{h}}(\underline{X})$ if a conditional test is used.

As noted in Section IIB, we may obtain the efficacy of such a conditional test by showing its asymptotic equivalence to a parametric test based on the same test statistic, not using a conditional structure. By applying this method it can be shown (as outlined in the Appendix) that the efficacy $E_{\underline{c},\underline{h}}$ of the (2k)-level sign detector is given by

$$E_{\underline{c},\underline{h}} = \frac{2\left\{\sum_{i=1}^{k} h_i[f(c_i) - f(c_{i-1})]\right\}^2}{\sum_{i=1}^{k} h_i^2[F(c_{i-1}) - F(c_i)]} \tag{29}$$

For a given vector $\underline{c}$ we can obtain the optimum value $\underline{h}(\underline{c})$ of $\underline{h}$, $\underline{h}(\underline{c}) = [h_1(\underline{c}), h_2(\underline{c}), \ldots, h_k(\underline{c})]$, maximizing the efficacy of equation (29). We have

$$h_i(\underline{c}) = \left[\frac{f(c_i) - f(c_{i-1})}{f(c_1)}\right]\left[\frac{1 - F(c_1)}{F(c_{i-1}) - F(c_i)}\right] \tag{30}$$

with the corresponding optimized value $E_{\underline{c}}$ of $E_{\underline{c},\underline{h}}$ obtained to be

$$E_{\underline{c}} = 2\sum_{i=1}^{k}\left\{\frac{[f(c_i) - f(c_{i-1})]^2}{F(c_{i-1}) - F(c_i)}\right\} \tag{31}$$

It should be noted at this point that the results of equations (29)-(31) hold not only for the conditional test based on $T_{\underline{c},\underline{h}}(\underline{X})$, but also for the parametric detector based on $T_{\underline{c},\underline{h}}(\underline{X})$. The results for the conditional test are derived from the asymptotic equivalence of the conditional and parametric (fixed-threshold) tests.

From equation (31), we find that as $k \to \infty$, if we consider increasingly finer quantizations of the real line, the efficacy $E_{\underline{c}}$ converges to the Fisher Information function $I(f)$ [13] defined by

$$I(f) = \int_{-\infty}^{\infty}\left[\frac{f'(x)}{f(x)}\right]^2 f(x)dx \tag{32}$$

(assuming this is finite and $f'(\cdot)$ exists). The Fisher Information function gives the efficacy of the locally optimum detector when $f(\cdot)$ is the noise density function. Thus the efficacy of the conditional nonparametric detector based on $T_{\underline{c},\underline{h}}(\underline{X})$, for optimum $\underline{h}$, can be very close to the efficacy of the locally optimum (parametric) detector when k is sufficiently large.

There is another interesting observation which can be made about the optimum parameters $\underline{h}(\underline{c})$ for the detector based on $T_{\underline{c},\underline{h}}(\underline{X})$. Assuming sufficient regularity, the locally optimum nonlinearity $\ell_f(\cdot)$ for H_1 versus K_1 when the noise density is $f(\cdot)$ is

$$\ell_f(x) = -f'(x)/f(x) \tag{33}$$

Suppose we look for the minimum mean-squared-error approximation $\ell_{\underline{c},\hat{\underline{h}}}(\cdot)$ of the nonlinearity $\ell_f(\cdot)$, for a given vector $\underline{c}$. The problem is therefore to minimize

$$E\{[\ell_{\underline{c},\underline{h}}(X) + \frac{f'(X)}{f(X)}]^2\}$$

$$= E\{\ell^2_{\underline{c},\underline{h}}(X) + 2\ell_{\underline{c},\underline{h}}(X)\,\frac{f'(X)}{f(X)}\} + I(f) \tag{34}$$

We have

$$E\{\ell^2_{c,h}(X) + 2\ell_{\underline{c},\underline{h}}(X)\,\frac{f'(X)}{f(X)}\} =$$

$$= 2\sum_{i=1}^{k} h_i^2 \int_{c_i}^{c_{i-1}} f(x)dx + 4\sum_{i=1}^{k} h_i \int_{c_i}^{c_{i-1}} f'(x)dx \tag{35}$$

Minimizing the mean-squared-error with respect to $\underline{h}$ is equivalent to minimizing the right side of equation (35) with respect to the h_i. By straightforward differentiation we find that the optimum values $\hat{h}_i$ are

$$\hat{h}_i = \frac{f(c_i) - f(c_{i-1})}{F(c_{i-1}) - F(c_i)} \tag{36}$$

and that the minimum mean-squared error, $mse_{\underline{c}}$, is

$$mse_{\underline{c}} = I(f) - E_{\underline{c}} \tag{37}$$

where $E_{\underline{c}}$ is defined by Equation (31).

We see, therefore, that the optimum values of the parameters $\underline{h}$ and $\underline{c}$ maximizing the efficacy of the detector are also the same as the values minimizing the mean-squared error between the locally optimum nonlinearity and its approximation $\ell_{\underline{c},\underline{h}}(\cdot)$. Actually, the values of $h_i(\underline{c})$ are scaled by an unimportant scale factor because of the initial assumption $h_1 = 1$.

There is another sense in which the conditional test based on $T_{\underline{c},\underline{h}}(\underline{X})$, with $\underline{h}$ given by equation (30), is an optimum test. Suppose in testing H_1 versus K_1 considerations of invariance or available data handling capabilities limit us to tests based on the N_{in} and N_{ip} , $i = 1,2,\ldots,k$, as defined by equation (28). For similar tests of size α (like the sign detector), the Lehmann-Scheffé theorem [24] allows attention to be restricted to tests conditioned on the statistic $\underline{N} = (N_1,N_2,\ldots,N_k)$ which is sufficient for H_1 and complete under H_1 . The conditional distribution of the N_{ip} is a product of independent binomial distributions where signal strength θ enters as a parameter. Thus the locally most powerful similar test for H_1 versus K_1 can be easily obtained as a conditional test [9]. Carrying out the operations to obtain the locally optimum test, one ends up with exactly the same test as that of the (2k)-level sign detector with the efficacy-optimum value of the vector $\underline{h}$ defined by equation (30). Of course, the locally optimum-similar test criterion is not an asymptotic (in sample size) criterion of performance.

For practical implementation perhaps the most useful detector to consider is the four-level sign detector obtained with $k = 2$. In this case the conditioning is only on N_1 (or N_2), since $N_1 + N_2 = n$, the sample size. For this four-level sign detector only two parameters $h_2 = h$ and $c_1 = c$ have to be set.

The structure of the four-level sign detector is shown in Figure 8. This structure is based on the fact that the four-level nonlinearity of this detector can be decomposed into a hard-limiter and a dead-zone limiter nonlinearity. For this detector, the table of thresholds and randomization probabilities is independent of c, but does depend on h. The entries in this table can be obtained easily from the binomial distribution function for the system parameters n and α.

It is possible to study both the asymptotic and finite-sample finite-signal performance of this detector. The asymptotic performance, in terms of efficacies, is obtained directly from equations (29)-(31). For Gaussian noise, the four-level sign detector with optimum parameters $c = c^*$ and $h = h^*$ has an efficacy E^* of $0.88/\sigma^2$, where σ^2 is the variance of the noise. The optimum values c^* and h^* in this case are 0.98σ and 0.30, respectively, so that h^* is independent of σ^2. Thus, the ARE of the four-level sign detector with respect to the linear detector for Gaussian noise is 0.88. This is a significant performance improvement over the case of the simple sign detector, which has an ARE with respect to the linear detector of 0.64 for Gaussian noise. The four-level sign detector also compares very favorably with the Wilcoxon detector which has an ARE with respect to the linear detector of 0.95 for Gaussian noise. It can also be shown [20] that the efficacy of this detector is not very sensitive to deviations in the value of the parameter c from the optimum. The parameter c can vary as much as 50 percent from its optimum value without degrading the performance appreciably. Efficacies showing excellent relative performance of the four-level sign detector can also be obtained for other noise densities as was done for the dead-zone limiter detector.

Since the conditional distribution of the test statistic of the four-level sign detector, given N_1, is based on the binomial distribution function, exact detection probabilities may be computed for finite samples and signals under specific alternatives. The computations are facilitated if we consider only values of h which can be written as ratios j/k for small values of integers j and k. In this case, further simplification is obtained if $j = 1$, implementation in such a case also being much simpler. Tables 1 and 2 show the power β for the four-level sign detector, for the detection problem defined by the hypotheses (H_1, K_1) with constant signal $(s_i = 1)$, with Gaussian noise of unit variance $(\sigma^2 = 1)$. These tables give β for different signal strengths θ, for different sample sizes n, and false alarm probabilities α. The power of the sign detector $(j = k; h = 1)$, the dead-zone limiter detector $(j = 0, k = 1)$, and the linear detector are also shown. The values of θ may be assumed to represent the signal-to-noise ratio θ/σ if the values of c are taken to represent the normalized parameter c/σ.

In these tables, the value 1/3 for h $(j = 1, k = 3)$ comes closest to the optimum value $h^* = 0.3$ for the Gaussian noise. The corresponding performance, for $c = c^* = 0.98$, is seen to be very good. Even for fairly large deviations of the parameter values from the optimum, the performance remains very good. These finite-sample, finite-signal, detection probabilities strengthen the previous conclusions from asymptotic comparisons about the excellent relative performance of the four-level sign detector.

As we have noted, for Gaussian noise the optimum value h^* of h for the four-level sign detector is independent of σ^2, the noise variance. However, c^* is proportional to σ. Thus even if Gaussian noise is assumed, its variance has to be estimated to set the value of c. The fact that the performance is not very sensitive to deviations from optimum values means that a rough estimate for c can be expected to give satisfactory results. In general, however, where noise cannot be assumed to be Gaussian or near-Gaussian, a different method has to be employed to obtain estimates of the optimum parameter values. It is always true, nevertheless, that nonparametric performance can be obtained with any values of c and h.

IV. TWO-INPUT CORRELATORS USING CONDITIONAL TESTING

The polarity-coincidence correlator (PCC) [30] is a device for the detection of a random signal common to two channels containing independent noises, a situation modeled as a test of the hypotheses H_2 versus K_2. It displays a nonparametric performance in the sense that the false alarm probability can be maintained at a constant value for noise processes with density functions having a zero median. The PCC is relatively easy to implement but, like the sign detector, its performance is rather poor compared to parametric detectors for noise densities which are Gaussian and near-Gaussian. We now describe some extensions of the PCC which are based on conditional tests and which have a considerably better overall detection performance relative to the PCC, without sacrifice of the nonparametric property or the simplicity of implementation.

The PCC uses the test statistic $T_P(\underline{X},\underline{Y})$ given by

$$T_P(\underline{X},\underline{Y}) = \sum_{i=1}^{n} \ell_0(X_i)\ell_0(Y_i) \tag{38}$$

where $\ell_0(x)$ is given by equation (11). The detector D_0 which is Neyman-Pearson optimal for testing H_2 versus K_2 under Gaussian statistics is based on the test statistic $T_0(\underline{X},\underline{Y})$:

$$T_0(\underline{X},\underline{Y}) = \sum_{i=1}^{n} (X_i + Y_i)^2 \tag{39}$$

A useful detector for H_2 versus K_2 is the correlator D_C which uses the test statistic $T_C(\underline{X},\underline{Y})$:

$$T_C(\underline{X},\underline{Y}) = \sum_{i=1}^{n} X_i Y_i \tag{40}$$

The asymptotic relative efficiency of the PCC with respect to D_0 for Gaussian noise [30] is only 0.2; with respect to D_C the PCC has an ARE for Gaussian noise [30] of 0.4. Although the PCC has an AREof 4.0 with respect to D_C for double-exponential noise densities [30], the rather poor performance for Gaussian and near-Gaussian noise is its major limiting aspect. For the same kind of reasons as those given in Section III, we now consider modified correlators which use a higher-level quantization of the data to obtain better performance.

Consider the test statistic $T_{DZC}(\underline{X},\underline{Y})$ defined by

$$T_{DZC}(\underline{X},\underline{Y}) = \sum_{i=1}^{n} \ell_c(X_i)\ell_c(Y_i) \tag{41}$$

where $\ell_c(\cdot)$ was defined in equation (12). Detector D_{DZC} based on $T_{DZC}(\cdot,\cdot)$ will be called the <u>dead-zone limiter coincidence detector</u>. It can be implemented as a nonparametric detector for H_2 versus K_2 when the noise density $f(\cdot)$ is symmetric about zero, by basing it on a conditional test exactly as we did for the dead-zone limiter detector in Section III. $T_{DZC}(\cdot,\cdot)$ may be represented as the sum of a function $f_c(\cdot,\cdot) = \ell_c(\cdot)\ell_c(\cdot)$ of each pair of observed values (X_i,Y_i). The function $f_c(X_i,Y_i)$ is illustrated in Figure 9. In general, $f_c(\cdot,\cdot)$ has the value zero for some of the n pairs of observations, and $T_{DZC}(\cdot,\cdot)$ has a distribution under H_2 which does depend on the density function $f(\cdot)$. Let N_{cp} and N_{cn} be the total numbers of pairs (X_i,Y_i) which fall in the regions labeled P and N in Figure 9, respectively, and define the random variable $N_c = N_{cp} + N_{cn}$. Then we can write

$$\begin{aligned} T_{DZC}(\underline{X},\underline{Y}) &= N_{cp} - N_{cn} \\ &= 2N_{cp} - N_c \end{aligned} \tag{42}$$

Although the exact distribution of N_{cp} under H_2 depends on $f(\cdot)$, <u>conditioned</u> on N_c the distribution of N_{cp} under H_2 is known, being binomial with parameters 1/2 (probability of success) and N_c (number of trials). This is true because of the symmetry of $f(\cdot)$. Thus a detector based on $T_{DZC}(\cdot,\cdot)$ for which the parameters of the threshold test are functions of the observations through N_c can result in nonparametric performance. Under K_2 more pairs of observations (X_i, Y_i) will fall in the regions labeled P in Figure 9 and $T_{DZC}(\underline{X},\underline{Y})$ will tend to have a larger value.

As we did for the dead-zone limiter detector, we can obtain the efficacy of the dead-zone limiter coincidence detector D_{DZC} from the mean and variance of $T_{DZC}(\underline{X},\underline{Y})$. The efficacy E_{DZC} of D_{DZC} may be easily shown to be

$$E_{DZC} = \frac{4f^4(c)}{[1 - F(c)]^2} \tag{43}$$

which is exactly the square of E_{DZ} given by equation (20). For Gaussian noise, the value c^* of c maximizing E_{DZC} is 0.612σ, when σ^2 is the noise variance, with a corresponding efficacy of $0.65/\sigma$. This results in an ARE of 1.62 with respect to the PCC, and an ARE of 0.65 with respect to the correlator D_C. These values indicate a significant improvement over the performance of the PCC.

For double-exponential noise densities, the PCC is, of course, the locally optimum detection scheme. Thus, when designed to give best performance for Gaussian noise, or any other noise density, D_{DZC} does not perform as well as the PCC for double-exponential noise densities. However, because the performance of D_{DZC} is not very sensitive to the value of c, we can use a value of c which is smaller than c^* for Gaussian noise. In this way we can still achieve significant performance improvements over the PCC for Gaussian noise, while maintaining superior performance with respect to the correlator D_C for double-exponential noise. For example, if a value of $c = 0.5c^*$ is used, the ARE of the dead-zone limiter coincidence detector with respect to the correlator D_C is 0.58 for Gaussian noise (still a significant improvement from 0.4 for the PCC) and is 1.71 for the case of the double-exponential noise density.

Thus D_{DZC} is an improvement over the PCC, giving much better detection performance for noise densities which are close to the Gaussian, while maintaining superior performance relative to D_C and D_0 for noise densities like the double-exponential density. It remains a simple detector in its implementation, the conditional test being easy to implement with a stored table of threshold and randomization probabilities.

Instead of using the function $f_c(\cdot,\cdot)$ of Figure 9 in forming the test statistic of a coincidence detector, we can consider the function $f_{c,h}(\cdot,\cdot)$, taking on values as indicated in Figure 10. The test statistic $T_{GC}(\cdot,\cdot)$ of the generalized coincidence correlator D_{GC} based on the function $f_{c,h}(\cdot,\cdot)$ is then given by

$$T_{GC}(\underline{X},\underline{Y}) = (2N_{1p} - N_1) + h(2N_{2p} - N_2) \tag{44}$$

Here N_{1p}, N_{2p}, N_{1n} and N_{2n} are the numbers of pairs of observations (X_i, Y_i) which fall in the regions labeled 1P, 2P, 1N, 2N, respectively, in Figure 10, with $N_1 = N_{1p} + N_{1n}$ and $N_2 = N_{2p} + N_{2n}$. The test statistic can also be written as

$$\begin{aligned} T_{GC}(\underline{X},\underline{Y}) &= \sum_{i=1}^{n} \ell_c(X_i)\ell_c(Y_i) + h\sum_{i=1}^{n} [\ell_0(X_i)\ell_0(Y_i) - \ell_c(X_i)\ell_c(Y_i)] \\ &= (1 - h)\sum_{i=1}^{n} \ell_c(X_i)\ell(Y_i) + h\sum_{i=1}^{n} \ell_0(X_i)\ell_0(Y_i) \end{aligned} \tag{45}$$

The coincidence detector based on $T_{GC}(\cdot,\cdot)$ is an extension of D_{DZC} just as the four-level sign detector is an extension of the dead-zone limiter detector. The distribution of $T_{GC}(\cdot,\cdot)$ conditioned on the value N_1 (and hence $N_2 = n - N_1$) is exactly known as a consequence of the assumption that the noise density function is symmetric. Thus a conditional nonparametric test can be constructed with $T_{GC}(\cdot,\cdot)$ as the test statistic.

The resulting detector D_{GC} will be more powerful that D_{DZC} if h is chosen appropriately; for example, the function $f_{c,h}(\cdot,\cdot)$ can be a better approximation than $f_c(\cdot,\cdot)$ to the function $f_R(X_i,Y_i) = X_iY_i$ of the correlator D_C .

It may be shown that the efficacy E_{GC} of D_{GC} is

$$E_{GC} = \frac{4\{f^2(c) + h[f^2(0) - f^2(c)]\}^2}{F^2(-c) + h^2[F^2(0) - F^2(-c)]} \tag{46}$$

with the optimum value $h(c)$ of h for a given c being

$$h(c) = \left[\frac{f^2(0) - f^2(c)}{f^2(c)}\right]\left[\frac{F^2(-c)}{F^2(0) - F^2(-c)}\right] \tag{47}$$

resulting in a maximum efficacy $E_{GC}(c)$ as a function of c which is

$$E_{GC}(c) = \frac{4f^4(c)}{[1 - F(c)]^2} + \frac{4[f^2(0) - f^2(c)]^2}{F^2(0) - F^2(-c)} \tag{48}$$

The efficacy $E_{GC}(c)$ can be maximized with respect to c . For Gaussian noise, the value c^* of c maximizing $E_{GC}(c)$ is 0.835σ , with a corresponding value E^*_{GC} of $E_{GC}(c)$ equal to $0.73/\sigma$, with σ^2 being the noise variance. For this value of c , $h(c) = h^*$ is 0.196.

In order to maintain good performance with respect to the correlator D_C for double-exponential-type noise densities, the value of c should be smaller than c^* (as in the previous case for D_{DZC}). The performance of D_{GC} for Gaussian noise is again not very sensitive to variations in the value of c from c^* so good performance for Gaussian noise can be obtained simultaneously. Similar performance characteristics can be established for non-Gaussian noise also.

The generalized coincidence detector we have been considering was based on a two-level (symmetric) partition of each quadrant of the 2-dimensional observation space. The detector can be extended to schemes which are based on k-level partitions, $k > 2$, of the observation-space quadrants. Further performance gains can be obtained from such nonparametric schemes, at the cost of increased complexity of implementation. In the limiting case of large k , the asymptotic performance can be made to approach that of the correlator D_C .

The coincidence detectors considered in this section give the same kind of performance improvements as those obtained by the generalized polarity coincidence correlators (GPCC) considered in [22]. However, for the GPCC, nonparametric behavior is obtained only if additional specific assumptions are made about the values of $F(c)$.

The extensions of the PCC considered in this section were analogous to the extensions of the sign detector considered in Section III. In Section III the detectors were analyzed in more detail and similar results may also, of course, be derived for the correlators of this section. Thus far we have applied conditional testing to obtain improvements of the simple polarity-based detection systems. In the next section we will describe an example of how conditional testing can be applied to advantage in detectors based on data ranking.

V. A CONDITIONAL WILCOXON DETECTOR

In the last two sections, we have analyzed some generalizations of detectors based only on polarities of received data. The motivation there was to obtain nonparametric detectors which remained simple to implement but which had better detection performances than the simple sign test and the PCC. Detectors based on the relative ranks of input samples in addition to their signs generally perform very well in comparison to parametric competitors and simple polarity-based detectors; on the other hand, the main factor limiting their usefulness is the problem of implementing the operation of ranking on a large number of observations. Feustel and Davisson [8] have suggested the use of "mixed" statistical tests to alleviate this implementation problem. We now consider the use of conditional tests to construct rank detectors simpler to implement than their standard-rank-test counterparts. The particular rank detector we will focus on is the one-input Wilcoxon detector; the procedure we will describe may also be applied in other rank-based detectors. We will also consider in this section a detector which is based on the Wilcoxon (one-input) test and which uses conditional testing as well as the technique of "mixed" statistical testing, giving a very simply implemented rank detector with the efficient performance associated with rank tests.

In order to test the null hypothesis H_1 against the alternative K_1, a Wilcoxon detector can be used to provide a constant-false alarm probability [28]. The Wilcoxon detector is an efficient fixed-threshold nonparametric detector and is based on a critical function defined in terms of a test statistic $T_W(\underline{X})$ given by

$$T_W(\underline{X}) = \sum_{i=1}^{n} \ell_0(X_i)R_i \tag{49}$$

where

$$R_i = \text{rank of } |X_i| \text{ in } \{|X_i|, |X_2|, \ldots, |X_n|\}$$

$\ell_0(\cdot)$ having been defined by equation (11).

From the received n-vector $\underline{X} = (X_1, X_2, \ldots, X_n)$ of independent components, suppose we form the sub-vector $\underline{X}^c = (X_1^c, X_2^c, \ldots, X_N^c)$ of N components, with $N \leq n$, such that $\underline{X}^c$ contains all those and only those components of $\underline{X}$ which are larger in absolute value than a non-negative constant c. The components X_i^c of the vector $\underline{X}^c$ are independent and under H_1, with the condition $|X_i^c| > c$, have a distribution which is symmetric about the origin. The Wilcoxon signed-rank test can therefore be applied to $\underline{X}^c$ for testing H_1 against K_1 with a constant false-alarm probability, because under H_1 the conditional distribution of the X_i^c is skewed to the right. Since N is a random variable the test is a conditional test, conditioned on the realization of N during any particular observation of the n-vector $\underline{X}$.

The advantage obtained is that N is less than or equal to n, so that on the average a smaller number of data values (depending on c) has to be ranked. The rule for selecting the sub-vector $\underline{X}^c$ could have been replaced by another probabilistic or deterministic rule; we will show that basing the detector test statistic on the sub-vector $\underline{X}^c$ obtained according to the simple scheme just described, with a suitably chosen constant c, can result in significant reduction of computational effort with little or almost no loss of detection performance over the case $c = 0$.

The detector based on the test statistic $T_W(\underline{X}^c)$ where $T_W(\cdot)$ is the signed rank statistic of equation (49) will be called the <u>conditional Wilcoxon detector</u>. For each of the $n + 1$ possible values of N, the thresholds and randomization probabilities required to obtain decisions with a false alarm probability α can be easily obtained from the null distribution of the Wilcoxon test statistic, from statistical tables or recursion formulas [13]. The table of $2(n + 1)$ values of thresholds and randomizations can be stored in a read-only memory (ROM) which forms part of the detector.

To get the efficacy of the conditional Wilcoxon detector, we need the expected value $E_\theta\{T_W(\underline{X}^c)\}$ of $T_W(\underline{X}^c)$ under the alternative of a constant positive signal θ, and its variance $\mathrm{Var}_0\{T_W(\underline{X}^c)\}$ under the hypothesis H_1. We first rewrite equation (49) with $\underline{X}^c$ in place of $\underline{X}$ in the form [3]

$$T_W(\underline{X}^c) = \sum_{i=1}^{N} \sum_{j=i}^{N} \ell_0(X_i^c + X_j^c) \tag{50}$$

This representation of $T_W(\underline{X}^c)$ makes it easier to obtain $E_\theta\{T_W(\underline{X}^c)\}$. If $f(x)$ and $F(x)$ are the density and distribution functions, respectively, of the noise, then the conditional density function $g_\theta(x)$ of the X_i^c (conditioned on $|X_i^c| > c$) under the alternative of a constant positive signal θ is given by

$$g_\theta(x) = \begin{cases} \dfrac{f(x - \theta)}{1 - F(c - \theta) + F(-c - \theta)}, & |x| > c \\ 0, & x \le c \end{cases} \tag{51}$$

From equations (50) and (51) we can get

$$E_\theta\{T_W(\underline{X}^c)\,|N\} = N\left[\frac{1 - F(c - \theta) - F(-c - \theta)}{1 - F(c - \theta) + F(-c - \theta)}\right] + \tfrac{1}{2} N(N - 1) \int_{-\infty}^{\infty}\int_{-\infty}^{\infty} \operatorname{sgn}(x + y) g_\theta(x) g_\theta(y)\,dx\,dy \tag{52}$$

We also have [13]

$$\operatorname{Var}_0\{T_W(\underline{X}_N^c)\,|N\} = \frac{1}{6} N(N + 1)(2N + 1) \tag{53}$$

Now as $n \to \infty$, the ratio $N/n \to [1 - F(c - \theta) + F(-c - \theta)]$ with probability one. Assuming that $F(c) < 1$, so that $T_W(\underline{X}^c)$ is asymptotically normally distributed, we obtain from equations (52) and (53) the result that the efficacy $E_W(c)$ of the conditional Wilcoxon detector is

$$E_W(c) = \frac{6[\int_c^\infty f^2(x)dx]^2}{[1 - F(c)]^3} \tag{54}$$

We can now base performance comparisons on this result. Note that when $c = 0$, equation (54) reduces to the efficacy of the fixed threshold Wilcoxon detector based on $T_W(\underline{X})$. From equation (54) we can obtain $ARE_W(c)$, the ARE of the conditional Wilcoxon detector with parameter c with respect to the fixed threshold Wilcoxon detector ($c = 0$) for a noise density $f(\cdot)$.

Figure 11 shows $ARE_W(c)$ as a function of c/σ for the important case where $f(\cdot)$ is a Gaussian noise density with variance σ^2. It is seen that the optimum value of c is 0.2σ resulting in an ARE of 1.01, representing a very slightly better performance from the conditional test. Also shown in Figure 11 is the fraction $E\{N\}/n$ under H_1 as a function of c/σ. Because of the flatness of the ARE curve as a function of c/σ, a value of $c = 0.5\sigma$ requires a ranking of only 0.62n samples on the average, while the ARE for this value of c is 0.98. The ARE curve of Figure 11 shows that the AREs are not very sensitive to the variations in c, so that a rough estimate of σ^2 for the purpose of picking c can result in good performance. Of course, the nonparametric performance of the detector is obtained for any non-negative value of c, because the threshold and randomization values depend only on N.

These results show the effectiveness of the conditional test in reducing the average requirements of a rank test which still performs almost as well, and in fact sometimes better, than the fixed-threshold Wilcoxon detector. Let us now consider applying the techniques of "mixed" tests [8] to further modify this conditional Wilcoxon rank test, to obtain good performance with a <u>fixed</u> substantial reduction in ranking requirement rather than obtaining only a reduced <u>average</u> computational requirement.

In [8], Feustel and Davisson introduced the technique of "mixed" statistical tests for reducing the implementation cost of nonparametric rank tests. For the Wilcoxon signed-rank detector, the "mixed" test statistic is formed by summing a set of p sub-statistics which are signed-rank statistics for groups of m independent input samples. The sample size is the product mp , and through this technique a ranking of mp values is replaced by p rankings of m values each. In [8] it is shown that even for rather small group sizes m the performance of the "mixed" test is quite close to that of the test based on a ranking of all mp samples. We will now show that further, significant reductions in ranking requirements can be obtained by using a <u>conditional</u> "mixed" Wilcoxon signed-rank test.

The conditional "mixed" Wilcoxon statistic which we study is obtained by considering groups of m samples from the sub-vector $\underline{X}^c$ instead of from the received vector $\underline{X}$. The number of groups P is now a random variable given by $P = [N/m]$. Since it has already been determined that basing a test statistic on $\underline{X}^c$ instead of on $\underline{X}$ can, in some cases, result in improved performance, it is not surprising that we can show that the conditional "mixed" Wilcoxon test allows us to obtain performance comparable to that of the "mixed" Wilcoxon test for smaller group sizes.

The test statistic of the conditional "mixed" Wilcoxon test with group size m is based on the statistic $T_{MW}(\underline{X}^c)$ given by

$$T_{MW}(\underline{X}^c) = \sum_{\ell=1}^{P} \sum_{i=1}^{m} \sum_{j=i}^{m} \operatorname{sgn}(X^c_{i,\ell} + X^c_{j,\ell}) \tag{55}$$

where

$$X^c_{i,\ell} = X^c_{i+(\ell-1)\cdot m} \quad , \quad P = \left[\frac{N}{m}\right]$$

The test based on $T_{MW}(\underline{X}^c)$ is a conditional test, since P is a random variable. Conditioned on the value of P , which can be any integer between 0 and n/m , the distribution of $T_{MW}(\underline{X}^c)$ under H_1 is completely determined. Thus a ROM device can be used to store the $2([n/m] + 1)$ values of thresholds and randomization probabilities required to obtain detection with a constant false alarm probability α . The memory requirement is reduced by a factor of $1/m$ by the grouping for the mixed test.

The efficacy of the conditional "mixed" Wilcoxon detector is obtained in the same way as that in which equation (54) was derived, using the fact that as $n \to \infty$, $P \to \infty$ almost surely. It can be shown easily that $E_{MW}(c,m)$, the efficacy of the conditional "mixed" Wilcoxon test with parameters c (determining $\underline{X}^c$) and m (group size), is given by

$$E_{MW}(c,m) = \frac{12\{f(c)[1 - F(c)] + (m - 1)\int_c^\infty f^2(x)dx\}^2}{(m + 1)(2m + 1)[1 - F(c)]^3} \quad (56)$$

When $m = 1$, we get

$$E_{MW}(c,1) = \frac{2f^2(c)}{1 - F(c)} \quad (57)$$

the efficacy of the dead-zone limiter detector considered in Section III. For $m \to \infty$, we have $E_{MW}(c,m) \to E_W(c)$ in accordance with what would be expected. When $c = 0$, the efficacy $E_{MW}(0,m)$ gives the efficacy of the "mixed" Wilcoxon test considered in [8]; with $f(\cdot)$ the Gaussian density, $E_{MW}(0,m)$ results in the same ARE expression as that found in [8].

For a given density $f(\cdot)$ and group size m , the parameter c may be picked to give the maximum value of the efficacy $E_{MW}(c,m)$. For the case of Gaussian noise, Figure 12 gives $E_{MW}(c^*,m)$ as a function of m , with the optimum value c^* of c for each value of m also indicated. Figure 12 also shows the values of $E_{MW}(0,m)$ for Gaussian noise. For the case $m = 1$, $E_{MW}(c^*,1)$ gives the efficacy of the optimized dead-zone limiter detector, and for both $c = 0$, $m = 1$, we get the efficacy of the simple sign detector. When $m \to \infty$, we get the efficacies $E_W(c)$ found previously.

Figure 12 leads to the interesting and useful conclusion that with just $m = 2$, the conditional "mixed" test with the optimum value $c^* = 0.6\sigma$ (when σ^2 is the variance of the noise) gives the same efficacy $(0.85/\sigma^2)$ as the "mixed" test with $m = 5$ when $c = 0$. A ranking of only two or three values at a time can be easily implemented, just as the memory required for a conditional test can be easily implemented, using ROMs. The combination of a "mixed" test and a conditional test thus results in a detector which has the attractive performance of a rank test without the requirement of ranking large numbers of data samples. Once again, as the curve for $c = 0.5\sigma$ in Figure 12 shows, the efficacies are relatively insensitive to changes in the values of c from the optimum, so that a rough estimate for σ^2 can lead to a value of c with close to the optimum performance. It is still true, in addition, that <u>any</u> value of c will retain nonparametric performance, since the threshold and randomization probabilities are functions of P only.

The efficacies $E_W(c)$ and $E_{MW}(c,m)$ can also be evaluated for non-Gaussian densities. For the class of generalized Gaussian densities of equation (9), the efficacies have been evaluated and are shown in Table 3. Efficacies are tabulated for different values of k ranging from 1.0 to 2.6, for $m = 2$ and $m \to \infty$, and for different values of c corresponding to the optimum value c^* for each k and to values which are optimum for Gaussian noise, including the case $c = 0$. These results show that the improvement from the conditional testing with $c > 0$ over the case $c = 0$ for $m = 2$ becomes more marked as k increases. For $m \to \infty$, corresponding to the ungrouped test, the efficacies of the conditional test change only a little from the case $c = 0$ for most densities in the class defined by equation (9); however, the average number of samples ranked relative to n decreases significantly as k increases.

Compared to the parametric linear detector which has an efficacy of $1/\sigma^2$, and is optimum for Gaussian noise, the nonparametric conditional "mixed" Wilcoxon detector with $m = 2$ is seen to be a useful alternative achieving a reasonably high relative efficiency with nonparametric performance for a wide range of densities in at least one class of density functions.

The analysis of the Wilcoxon detector modified through conditional testing shows that conditional testing may be used effectively to reduce the complexity of implementation of many other rank tests. Although it does introduce the need for a table-look-up procedure to be implemented, the ranking requirement this replaces results in a system which is usually simpler to implement. In the case of most rank tests, of course, it is the implementation problem that limits their usefulness.

VI. AN APPLICATION OF CONDITIONAL TESTING IN NONPARAMETRIC DETECTION WITH DEPENDENT DATA

In almost all analyses of nonparametric detectors (including those considered thusfar in this chapter), it is assumed that the input data is a sequence of independent random variables. If this assumption is violated, it is well known that the detectors lose their nonparametric performance [6,29]. When the received information is a continuous-time random process, this assumption limits the sampling rate which can be used in obtaining sampled data. In [18], modifications to standard nonparametric detectors are suggested which can be implemented to obtain nonparametric performance under conditions of dependent inputs. In this section we focus on a particular modified detector to illustrate one way in which conditional testing can be useful in the construction of nonparametric detectors for dependent inputs.

Suppose that the vector of n observations $\underline{X}$ is obtained by sampling a continuous-time process. If we assume that this process is T-dependent with a normalized autocorrelation function $\rho(\tau)$ which has support $(-T,T)$, then in general a sampling rate of γ/T will result in a $(\gamma-1)$-dependent sequence of samples when we assume γ to be a positive integer. [A process is T-dependent if samples separated by a time interval larger than or equal to T are independent. This assumption makes the analysis tractable; it can be shown [18] that this assumption may be relaxed.] We can derive a sequence of independent multivariate observations from the sequence of $(\gamma-1)$-dependent samples by considering groups of m consecutive samples, with $(\gamma - 1)$ samples skipped between groups. Denote the ith group of m samples by $\underline{Z}_i$. We then consider a multivariate-to-univariate transformation applied to each $\underline{Z}_i$, converting it to a reduced observation Z^i. We thus obtain a sequence of independent variables Z^i.

In [18], several detectors operating on independent sequences of data derived from dependent samples are analyzed. Let us concentrate on one simple detector, the sign detector. If N is the number of independent samples which can be obtained in a given observation interval from the continuous-time input process with a sampling rate $1/T$, then a γ-fold increase in the sampling rate yields γ as many samples $n = N\gamma$. By using the grouping and skipping method followed by transformation to the Z^i, a total of $N\gamma m/(m + \gamma - 1)$ samples are utilized for detection. If the transformation applied to obtain the Z^i is properly chosen, an improvement in detector performance can be expected. Consider the case where any k-variate joint density $f(x_1,x_2,\ldots,x_k)$ of the noise process is diagonally symmetric; that is

$$f(x_1,x_2,\ldots,x_k) = f(-x_1,-x_2,\ldots,-x_k) \tag{58}$$

Then a transformation from $\underline{Z}_i$ to Z^i defined as a summing of the m components of $\underline{Z}_i$ is an appropriate, simple transformation for detecting a constant positive signal which is additive on the noise. Under the null noise-only hypothesis, Z^i has a symmetric distribution which gets shifted to the right when signal is present. Thus the sign test can be performed on the $\{Z^i\}$ instead of on $\underline{X}$. It is shown in [18] that the resulting detector is more efficient than the nonparametric sign detector operating only on directly obtained independent input samples. The appropriate asymptotic relative performance index here is the asymptotic relative processing time (ARPT) [18], which is a comparison of lengths of observation times rather than numbers of samples required by two detectors achieving the same performance asymptotically. The ARPT of this modified sign detector relative to the standard sign detector has been obtained in [18] and has been shown to have values significantly larger than unity in many cases of interest.

The use of a summing operation diminishes one of the attractive features of a sign detector, namely extreme simplicity of implementation. In addition, the summing operation was not arrived at as an optimal transformation, so we are led to consider other, nonlinear, transformations. One possibility which suggests itself is to sum the <u>signs</u> of the m components of $\underline{Z}_i$ to effect the transformation. Consider in particular the important case when $m = 2$ and $\gamma = 2$. Then Z^i can be either $+2$, -2, or 0. Under the null hypothesis, Z^i has equal probability of being $+2$ or -2, but in addition has a non-zero probability of having the value 0. Thus the distribution of the modified sign detector test statistic will depend on the noise process. At this point we turn to conditional testing. It is clear that <u>conditioned</u> on the total number of non-zero Z^i, the distribution of the modified sign detector test statistic is a known shifted binomial distribution. Thus it is possible to obtain nonparametric detection with this transformation scheme by using a conditional test.

It has been shown in [18] that the use of this nonlinear transformation results in even better performance of the modified nonparametric sign detector for dependent data. This modified sign detector for dependent data based on conditional testing illustrates the usefulness of the conditioning principle in nonparametric detector formulations for dependent input data.

The conditional test considered here for nonparametric detection with dependent data (for the special case $\gamma = 2$, $m = 2$) has been used by Chatterjee [4] as an extension of the sign test for location in the bivariate case. In [4], the test is justified more formally using the concepts of invariance and sufficiency and its properties are studied in more detail. In addition to describing Chatterjee's test, [27] discusses other procedures for nonparametric conditional testing based on multivariate data. Some of these procedures appear to have interesting applications to combat the dependent samples problem of nonparametric detection.

VII. CONCLUDING REMARKS

In this chapter the concept of conditional testing was applied to generalize and extend several well-known nonparametric detection techniques. In general, the resulting detectors display very good performances and remain quite simple to implement with hardware commonly available. The analyses and results presented here indicate quite clearly the advantages which may be obtained from conditional structures for nonparametric detection. In addition, further research in this direction can be expected to provide results adding considerably to the usefulness of such detection methods.

One important area of further research is the investigation of methods to obtain simple, effective estimates of the optimal parameters $\underline{c}$ and $\underline{h}$ of the input quantizers. For Gaussian noise, for example, the optimum value of $\underline{c}$ depends on the variance which may therefore be estimated by a recursive algorithm. In general, algorithms to obtain rough estimates of the type of the noise distribution may be implemented. The use of decision-directed adaptation and stochastic-approximation techniques to update estimates of the quantizer parameters could prove useful.

Other areas of investigation include an analysis of sequential detectors using conditional testing and the extension of detectors considered here to include multi-input detectors. For the problem of dependent samples, conditional testing appears to have the potential of providing direct methods of performing nonparametric detection. In [27] some conditional multivariate tests are described which may be adapted to the signal detection problem. For appropriate noise models such as the Markov model, conditional testing may turn out to provide natural nonparametric detection schemes.

APPENDIX. EFFICACY OF THE (2k)-LEVEL SIGN DETECTOR

We are considering conditional tests with test functions $\Phi(\underline{X})$ of the form

$$\Phi(\underline{X}) = \begin{cases} 1 & , T_n(\underline{X}) > t_{n,\alpha}(\underline{X}) \\ r_{n,\alpha}(\underline{X}) & , T_n(\underline{X}) = t_{n,\alpha}(\underline{X}) \\ 0 & , T_n(\underline{X}) < t_{n,\alpha}(\underline{X}) \end{cases} \tag{59}$$

for testing H_1 versus K_1 at size α. The functions $r_{n,\alpha}(\underline{X})$ and $t_{n,\alpha}(\underline{X})$ are functions of $\underline{X}$ through a statistic $S_n(\underline{X})$. In our case we may define a normalized statistic $T_n(\underline{X})$ as

$$T_n(\underline{X}) = \frac{1}{\sqrt{n}} T_{\underline{c},\underline{h}}(\underline{X}) \tag{60}$$

where $T_{\underline{c},\underline{h}}(\underline{X})$ was defined in equation (27), with $S_n(\underline{X})$ the vector $\underline{N} = (N_1, N_2, \ldots, N_k)$. The subscripts n emphasize dependence on the sample size.

Following the notation in [9, chapter 7], we define a conditional distribution function $F_n(s;\underline{x})$ as

$$F_n(s;\underline{x}) = P_{H_1}\{T_n(\underline{X}) \leq s \,|\, S_n(\underline{X}) = S_n(\underline{x})\} \tag{61}$$

Let $\Phi^*(\underline{X})$ be the test function of the parametric test of size α based on $T_n(\underline{X})$, obtained by choosing appropriate constant values $t_{n,\alpha}$ and $r_{n,\alpha}$ for the threshold and randomization probability, respectively, in equation (59). Suppose we consider a sequence $\{\theta_n\}$ of values of θ in equation (7), defining K_1 , where we also restrict the signal samples s_i to be all equal to unity. Then the following two theorems due to Hoeffding [15] can be used to establish the asymptotic power equivalence of the tests based on $\Phi(\underline{X})$ and $\Phi^*(\underline{X})$ for testing H_1 versus K_1 .

Theorem 1: For the sequence of distributions $\{\theta_n\}$ for $\underline{X}$, if $F_n(s;\underline{X})$ converges in probability to a distribution function $G(s)$ at every point of continuity of $G(s)$, if $G(s) = 1 - \alpha$ has a unique solution $s = t_\alpha$, a point of continuity of $G(s)$, then $t_{n,\alpha}(\underline{X})$ converges in probability to t_α . And, if there exists a function $H(s)$ continuous at t_α such that

$$P_{\theta_n}\{T_n(\underline{X}) \leq s\} \to H(s)$$

at every continuity point of $H(s)$, then, for the sequence of distributions $\{\theta_n\}$, the power of $\Phi(\underline{X})$ converges to $1 - H(t_\alpha)$. Also, if $t_{n,\alpha}$ converges to t_α , then for the sequence $\{\theta_n\}$, the power of $\Phi^*(\underline{x})$ converges also to $1 - H(t_\alpha)$.

The proof can be found in [15]. To check whether $F_n(s;\underline{X})$ converges to a distribution $G(s)$, another theorem due to Hoeffding may be used. If we define a random variable $C_n(\underline{x})$ as one having the distribution function $F_n(s;\underline{x})$ [i.e., $P\{C_n(\underline{x}) \leq s\} = F_n(s,\underline{x})$], then we have

Theorem 2: A necessary and sufficient condition that $F_n(s;\underline{X})$ converges in probability to $G(s)$ is that

$$P\{C_n(\underline{X}) \leq s\} \to G(s)$$

and

$$P\{C_n(\underline{X}) \leq s, C_n'(\underline{X}) \leq s\} \to G^2(s)$$

where $C_n(\underline{x})$ and $C_n'(\underline{x})$ are independent and identically distributed random variables, each with distribution function $F_n(s;\underline{x})$.

The proof of this theorem is also given in [15].

After the asymptotic power equivalence of $\Phi(\underline{X})$ and $\Phi^*(\underline{X})$ is established, the efficacy of the test based on $\Phi(\underline{X})$ can be obtained as the efficacy of the parametric test based on $\Phi^*(\underline{X})$. To illustrate how the preceding theorems may be employed in our general case, we consider the special case $k = 2$ (four-level sign detector) and outline the derivation of the efficacy. This derivation may then be easily generalized to hold for the case $k > 2$.

The conditional distribution $F_n(s;\underline{x})$ can be written as $F_n(s;n_1)$ (n_1 being a realization of N_1) in the case we are considering. To prove that this converges to a zero-mean Gaussian distribution by Theorem 2, we consider the joint distribution of $C_n(\underline{X})$ and $C_n'(\underline{X})$ defined as

$$C_n(\underline{X}) = \frac{1}{\sqrt{n}} \sum_{i=1}^{N_1} U_i + \frac{h}{\sqrt{n}} \sum_{i=1}^{n-N_1} V_i \tag{62}$$

$$C_n'(\underline{X}) = \frac{1}{\sqrt{n}} \sum_{i=1}^{N_1} U_i' + \frac{h}{\sqrt{n}} \sum_{i=1}^{n-N_1} V_i' \tag{63}$$

$\{U_i\}$, $\{U_i'\}$, $\{V_i\}$, and $\{V_i'\}$ being sequences of independent random variables taking on values -1 and +1 with equal probability.

Let $m(w_1,w_2;n)$ be the moment generating function (mgf) of the joint distribution of the random variables $C_n(\underline{X})$ and $C_n'(\underline{X})$. Under the alternative θ_n, let p_n be the probability that X_i lies in $[c,\infty)$ or $(-\infty,-c)$, and let q_n be the probability that X_i lies in $[-c,c)$. Define the quantities p_n, q_n by

$$p_n = 2p + \beta_n - \gamma_n \tag{64}$$

$$q_n = 1 - 2p - (\beta_n - \gamma_n) \tag{65}$$

where

$$1 - F(c - \theta_n) = 1 - F(c) + \beta_n = p + \beta_n \tag{66}$$

and

$$F(-c - \theta_n) = F(-c) - \gamma_n = p - \gamma_n \tag{67}$$

We assume that $\beta_n - \gamma_n \to 0$, so that the sequence of alternatives approaches the hypothesis. Now it can be shown easily that

$$\begin{aligned} m(w_1,w_2;n) = \frac{1}{4^n}\{&p_n[\exp(w_1/\sqrt{n}) + \exp(-w_1/\sqrt{n})][\exp(w_2/\sqrt{n}) + \exp(-w_2/\sqrt{n})] \\ &+ q_n[\exp(w_1h/\sqrt{n}) + \exp(-w_1h/\sqrt{n})][\exp(w_2h/\sqrt{n}) + \exp(-w_2h/\sqrt{n})]\}^n \end{aligned} \tag{68}$$

This mgf follows from the fact that N_1 has a binomial distribution and can be obtained by first conditioning on N_1. From this we conclude that

$$\lim_{n\to\infty} m(w_1,w_2;n) = \exp\{\frac{w_1^2}{2}[2p + h^2(1 - 2p)]\}\exp\{\frac{w_2^2}{2}[2p + h^2(1 - 2p)]\} \tag{69}$$

Thus, by Theorem 2, $F_n(s;N_1)$ converges to the Gaussian distribution function with zero mean and variance $2p + h^2(1 - 2p)$, under any sequence of alternatives for which p_n as defined by equation (64) approaches $2p$ of equation (66).

To show that $T_n(\underline{X})$ has also unconditionally an asymptotically Gaussian distribution with variance $2p + h^2(1 - 2p)$ under the hypothesis, we write $T_n(\underline{X})$ as

$$T_n(\underline{X}) = \frac{1}{\sqrt{n}} \sum_{i=1}^{N_1} V_i + \frac{h}{\sqrt{n}} \sum_{i=1}^{n-N_1} V_i'$$

It follows directly from the result just proved that this has the stated distribution as $n \to \infty$. This proves the convergence of $t_{n,\alpha}$ to t_α (Theorem 1), so that the conditional and unconditional tests are asymptotically power equivalent.

By using equation (4), the efficacy of the test based on Φ^* can easily be derived as $E_{\underline{c},\underline{h}}$ of equation (29) for k = 2 with $c_1 = c$ and $h_2 = h$. This is also, therefore, the efficacy of the conditional four-level sign detector. For the general case when $k > 2$, the procedure for obtaining $E_{\underline{c},\underline{h}}$ is a direct extension of the procedure followed above. Equations (62) and (63) are now composed of k terms and the mgf of equation (68) is obtained by first conditioning on the vector $\underline{N}$ which has a multinomial distribution.

REFERENCES

1. Brownstein, H. and L. Kurz (1970) "Parametric and nonparametric detectors based on a sequential sample median test," Inf. and Control 17: 417-435.
2. Capon, J (1959) "Nonparametric methods for the detection of signals in noise," Tech. Report T-1/N, Columbia University, New York, N.Y.
3. Carlyle, J.W. (1968) "Nonparametric methods in detection theory," in Communication Theory (A.V. Balakrishnan, ed.), McGraw-Hill: New York.
4. Chatterjee, S.K. (1966) "A bivariate sign test for location," Ann. Math. Stat. 37: 1771-1782.

5. Ching, Y.C. and L. Kurz (1972) "Nonparametric detectors based on m-interval partitioning," IEEE Trans. Inf. Th. IT-18: 251-257.

6. Davisson, L.D., E.A. Feustel and J.W. Modestino (1970) "The effects of dependence on nonparametric detection," IEEE Trans. Inf. Th. IT-16: 32-41.

7. Feller, W. (1971) An Introduction to Probability Theory and Its Applications, Vol. II. J. Wiley & Sons: New York.

8. Feustel, E.A. and L.D. Davisson (1967) "The asymptotic relative efficiency of mixed statistical tests," IEEE Trans. Inf. Th. IT-13: 247-255.

9. Fraser, D.A.S. (1957) Nonparametric Methods in Statistics. J. Wiley & Sons: New York.

10. Gastwirth, J.L. (1966) "On robust procedures," J. of the Am. Stat. Assoc. 61: 929-948.

11. Groeneveld, R.A. (1972) "Asymptotically optimal group rank tests for location," J. Am. Stat. Assoc. 67: 847-849.

12. Groeneveld, R.A. (1969) "The efficiency of level tests in the detection of weak stochastic signals," IEEE Trans. Inf. Th. IT-15: 502-504.

13. Hajek, J. and Z. Sidak (1967) Theory of Rank Tests. Academic: New York.

14. Hodges, J.L. and E.L. Lehmann (1956) "The efficiency of some nonparametric competitors of the t-test," Ann. Math. Stat. 27: 324-335.

15. Hoeffding, W. (1952) "The large sample power of tests based on permutations of observations," Ann. Math. Stat. 23: 169-192.

16. Johnson, N.L. and S. Kotz (1970) Continuous Univariate Distributions -- I, II. Houghton Mifflin: Boston.

17. Kanefsky, M. and J.B. Thomas (1965) "On polarity detection schemes with non-Gaussian inputs," J. Franklin Inst. 280: 120-138.

18. Kassam, S.A. and J.B. Thomas (1975) "A class of nonparametric detec-

tors for dependent input data," IEEE Trans. Inf. Th. IT-21: 431-437.

19. Kassam, S.A. and J.B. Thomas (1976) "The dead-zone limiter: An application of conditional tests in nonparametric detection," J. of the Acous. Soc. Am. 60: 857-862.

20. Kassam, S.A. and J.B. Thomas (1976) "Generalizations of the sign detector based on conditional tests," IEEE Trans. Commun. COM-24: 481-487.

21. Kassam, S.A. and J.B. Thomas (1975) "A modified sign detector based on a four-level input quantization," Proc. 13th Ann. Allerton Conf. on Circuit and System Theory, 204-210.

22. Lainiotis, D.G. (1968) "On a class of generalized polarity coincidence correlators," IEEE Trans. Inf. Th. IT-14: 603-604.

23. Lehmann, E.L. (1959) Testing Statistical Hypotheses. J. Wiley & Sons: New York.

24. Lehmann, E.L. and H. Scheffé (1950) "Completeness, similar regions, and unbiased estimation -- part 1," Sankhyā 10: 305-340.

25. Miller, J.H. and J.B. Thomas (1972) "Detectors for discrete-time signals in non-Gaussian noise," IEEE Trans. Inf. Th. IT-18: 241-250.

26. Noether, G.E. (1955) "On a theorem of Pitman," Ann. Math. Stat. 26: 64-68.

27. Puri, M.L. and P.K. Sen (1971) Nonparametric Methods in Multivariate Analysis. J. Wiley & Sons: New York.

28. Thomas, J.B. (1970) "Nonparametric detection," Proc. IEEE 58: 623-631.

29. Wolff, S.S., J.L. Gastwirth, and H. Rubin (1967) "The effect of autoregressive dependence on a nonparametric test," IEEE Trans. Inf. Th. IT-13: 311-313.

30. Wolff, S.S., J.B. Thomas and T.R. Williams (1962) "The polarity-coincidence correlator: A nonparametric detection device," IRE Trans. Inf. Th. IT-18: 5-9.

		θ, signal amplitude				
		0.0	0.1	0.2	0.3	0.4
	j=1, k=2	0.010	0.045	0.141	0.325	0.565
	j=1, k=3	0.010	0.046	0.147	0.339	0.585
c=0.61	j=1, k=4	0.010	0.046	0.148	0.341	0.588
	j=1, k=5	0.010	0.046	0.148	0.341	0.588
	j=0, k=1	0.010	0.044	0.139	0.320	0.558
	j=1, k=2	0.010	0.045	0.145	0.335	0.580
	j=1, k=3	0.010	0.047	0.152	0.351	0.603
c=0.98	j=1, k=4	0.010	0.047	0.152	0.351	0.603
	j=1, k=5	0.010	0.046	0.150	0.346	0.596
	j=0, k=1	0.010	0.041	0.124	0.285	0.508
	j=1, k=2	0.010	0.045	0.143	0.330	0.573
	j=1, k=3	0.010	0.046	0.149	0.346	0.597
c=1.20	j=1, k=4	0.010	0.046	0.148	0.344	0.594
	j=1, k=5	0.010	0.045	0.145	0.336	0.583
	j=0, k=1	0.010	0.037	0.107	0.242	0.440
Sign Detector j=k		0.010	0.038	0.112	0.252	0.452
Linear Detector		0.010	0.053	0.181	0.419	0.692

Table 1 Powers for the 4-level sign, the dead-zone limiter, the sign and the linear detectors; Gaussian noise of unit variance, sample size $n = 50$, $\alpha = 0.01$.

		θ, signal amplitude				
		0.0	0.1	0.2	0.3	0.4
	j=1, k=2	0.001	0.014	0.094	0.331	0.670
	j=1, k=3	0.001	0.014	0.100	0.351	0.695
c=0.61	j=1, k=4	0.001	0.015	0.101	0.354	0.700
	j=1, k=5	0.001	0.015	0.101	0.353	0.700
	j=0, k=1	0.001	0.014	0.092	0.326	0.664
	j=1, k=2	0.001	0.014	0.098	0.346	0.691
	j=1, k=3	0.001	0.015	0.106	0.369	0.719
c=0.98	j=1, k=4	0.001	0.015	0.105	0.369	0.719
	j=1, k=5	0.001	0.015	0.103	0.362	0.712
	j=0, k=1	0.001	0.012	0.078	0.281	0.602
	j=1, k=2	0.001	0.014	0.096	0.340	0.683
	j=1, k=3	0.001	0.015	0.103	0.363	0.712
c=1.20	j=1, k=4	0.001	0.015	0.102	0.359	0.708
	j=1, k=5	0.001	0.014	0.099	0.348	0.694
	j=0, k=1	0.001	0.010	0.062	0.224	0.510
Sign Detector j=k		0.001	0.011	0.065	0.232	0.518
Linear Detector		0.001	0.018	0.135	0.464	0.819

Table 2 Powers for the 4-level sign, the dead-zone limiter, the sign and the linear detectors; Gaussian noise of unit variance, sample size $n = 100$, $\alpha = 0.001$.

	k, Parameter of Generalized Gaussian Density	1.0	1.4	1.8	2.0	2.2	2.6
m=2	c^*/σ	0	0.06	0.35	0.50	0.63	0.85
	$\sigma^2 E_{MW}(c^*,m)$	1.800	1.055	0.873	0.846	0.839	0.857
	$\sigma^2 E_{MW}(0.5\sigma,m)$	0.893	0.891	0.862	0.846	0.831	0.804
	$\sigma^2 E_{MW}(0,m)$	1.800	1.045	0.805	0.742	0.697	0.638
$m\to\infty$	c^*/σ	0	0.00	0.06	0.20	0.34	0.58
	$\sigma^2 E_W(c^*)$	1.500	1.124	0.992	0.968	0.964	0.991
	$\sigma^2 E_W(0.2\sigma)$	1.131	1.034	0.983	0.968	0.958	0.946
	$\sigma^2 E_W(0)$	1.500	1.124	0.990	0.955	0.930	0.900
	$E\{N\}/n$ under H_1 for optimum c	1.00	1.00	0.95	0.84	0.74	0.59

Table 3 Optimum values of c , efficacies relative to $1/\sigma^2$ and average fraction of samples ranked under H_1 for the conditional Wilcoxon detectors for generalized Gaussian noise.

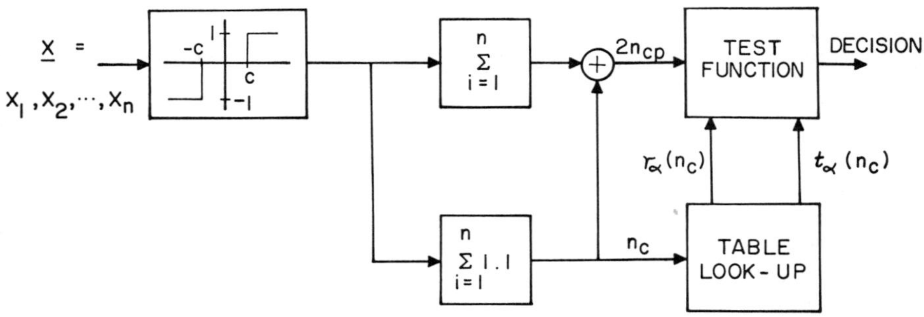

Figure 1 The dead-zone limiter detector.

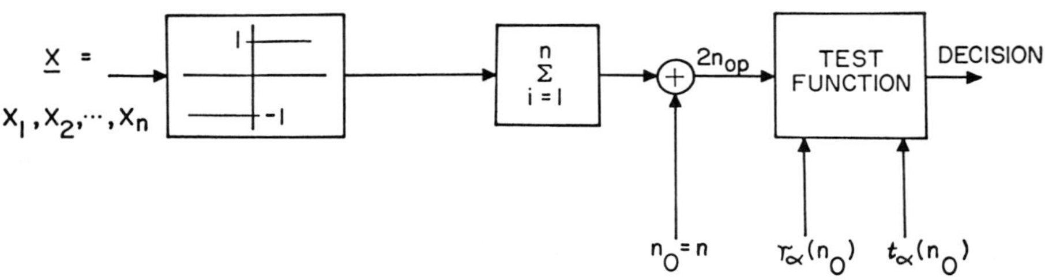

Figure 2 The sign detector.

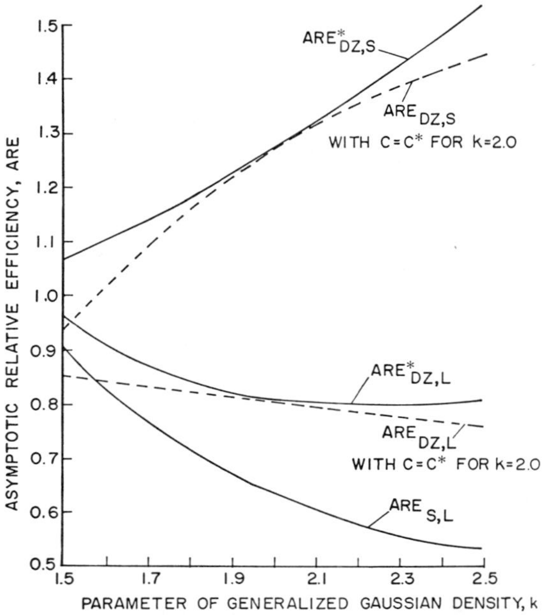

Figure 3 Asymptotic performance of the dead-zone limiter detector for generalized Gaussian densities.

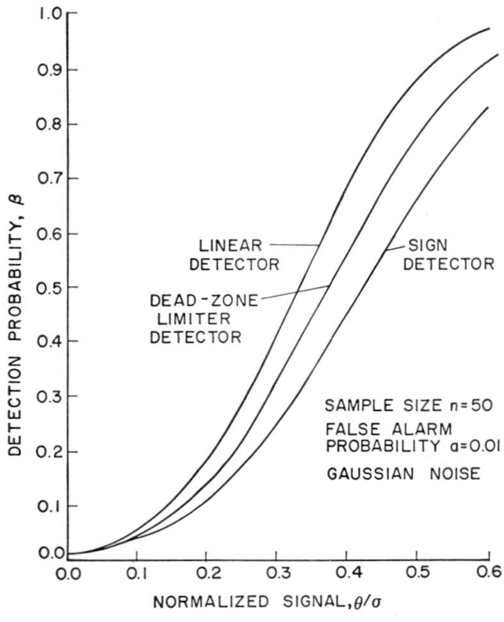

Figure 4 Finite-sample performance of the dead-zone limiter detector; $n = 50$, Gaussian noise with variance σ^2.

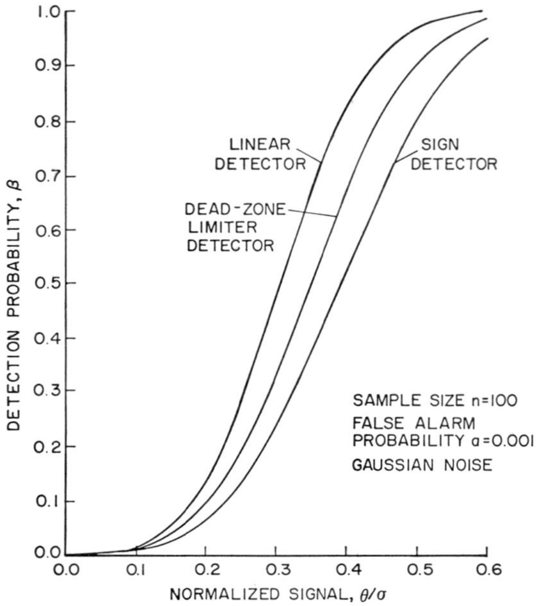

Figure 5 Finite-sample performance of the dead-zone limiter detector; $n = 100$, Gaussian noise with variance σ^2.

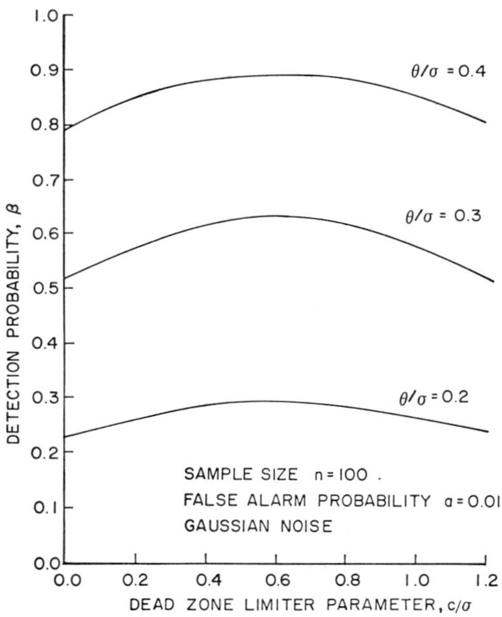

Figure 6 Variation of finite-sample performance with c of the dead-zone limiter detector; $n = 100$, Gaussian noise with variance σ^2.

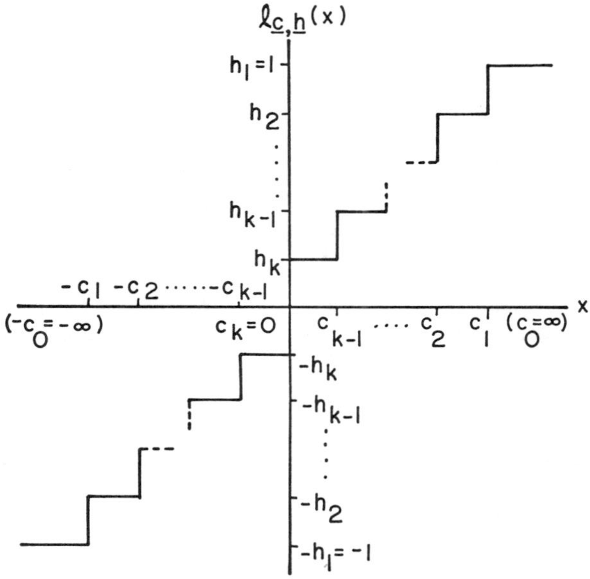

Figure 7 The (2k)-level quantizer.

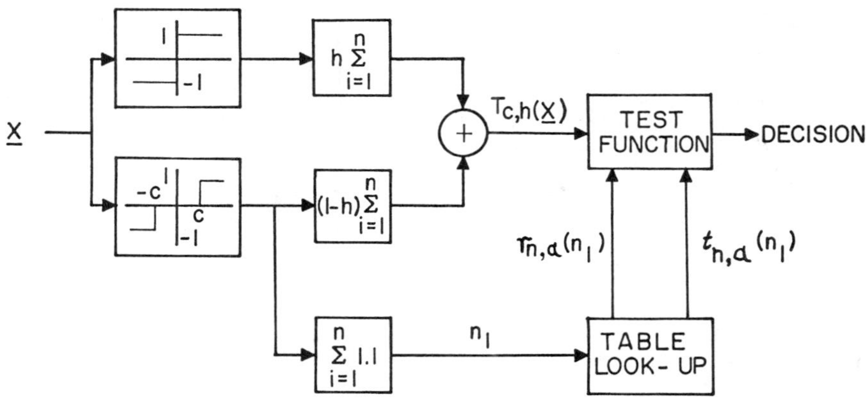

Figure 8 The 4-level sign detector.

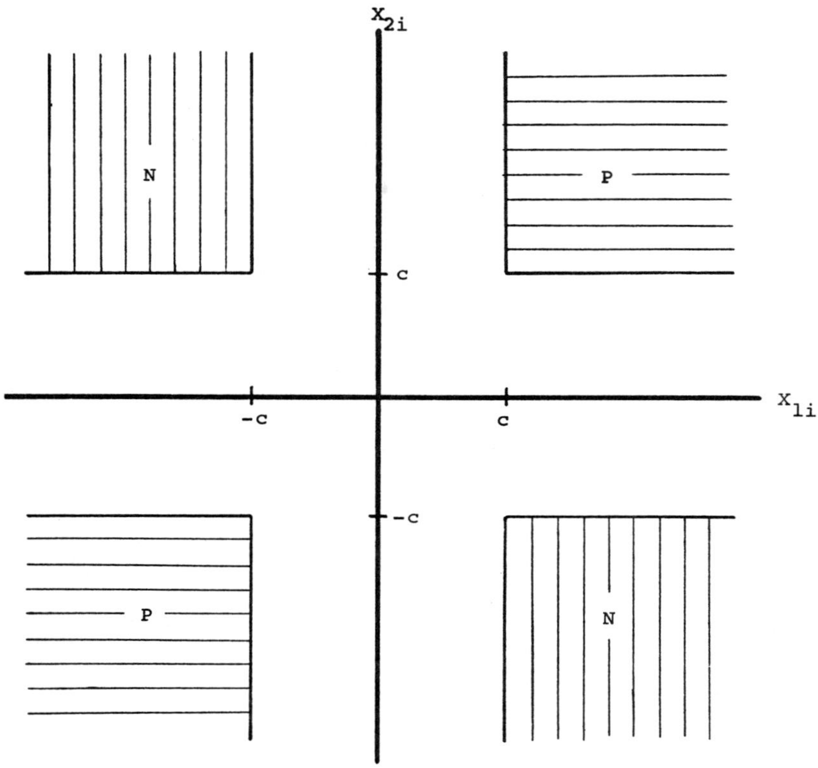

Figure 9 The function $f_c(X_i,Y_i)$. $f_c(X_i,Y_i) = \begin{cases} 1, & (X_i,Y_i) \in P \\ -1, & (X_i,Y_i) \in N \\ 0, & \text{otherwise} \end{cases}$

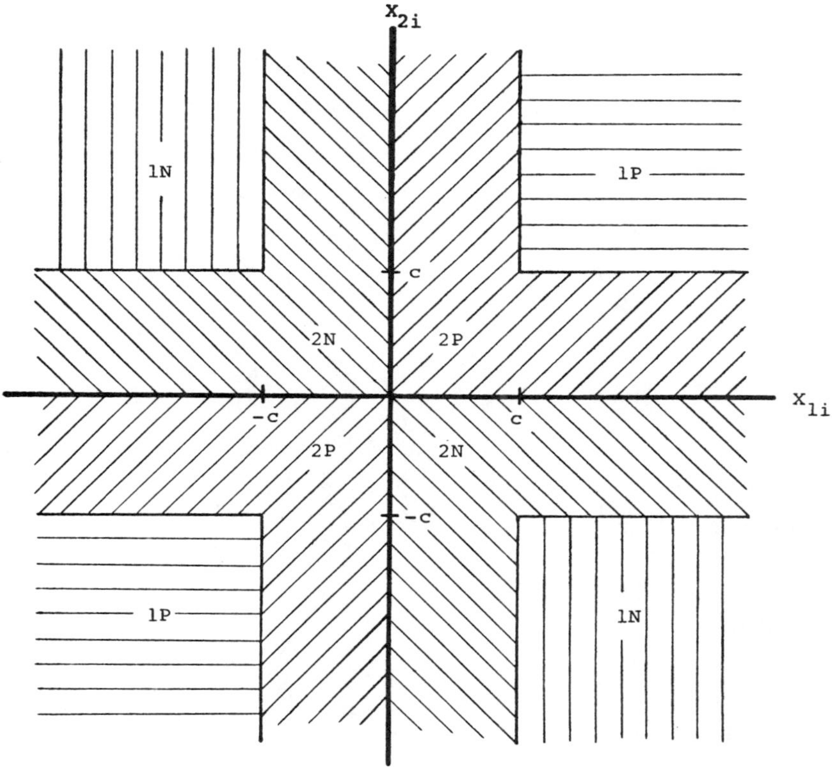

Figure 10 The function $f_{c,h}(X_i,Y_i)$. $f_{c,h}(X_i,Y_i) = \begin{cases} 1, & (X_i,Y_i) \in 1P \\ h, & (X_i,Y_i) \in 2P \\ -h, & (X_i,Y_i) \in 2N \\ -1, & (X_i,Y_i) \in 1N \end{cases}$

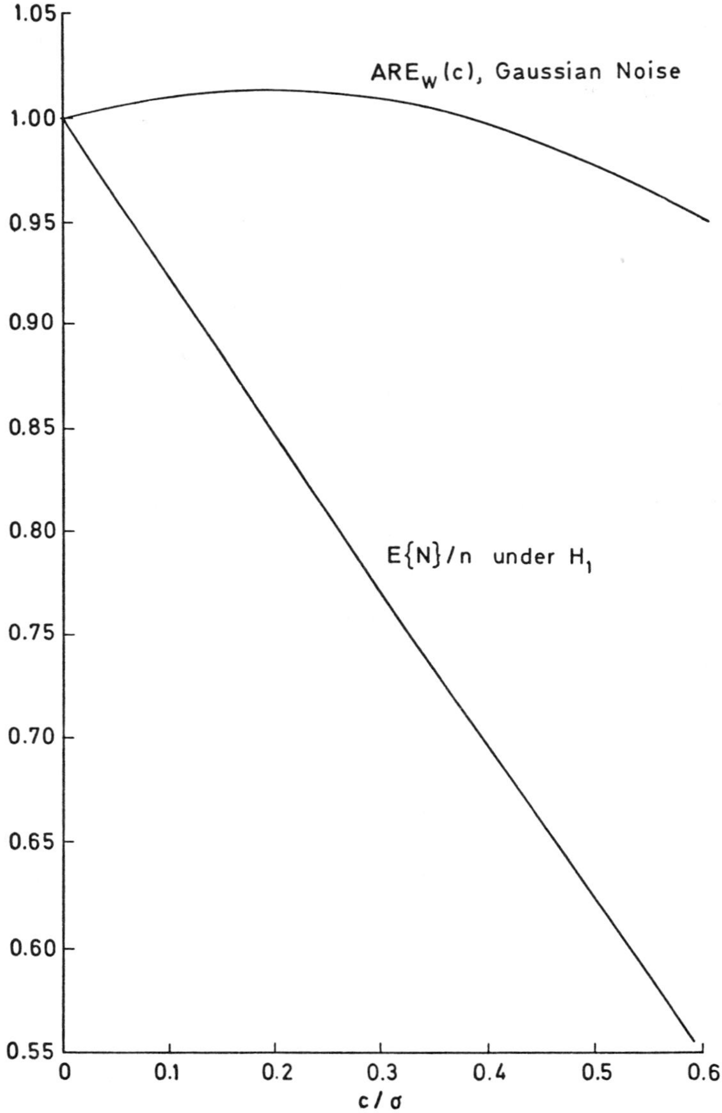

Figure 11 ARE of the conditional Wilcoxon detector relative to the fixed threshold Wilcoxon detector, Gaussian noise, variance σ^2 .

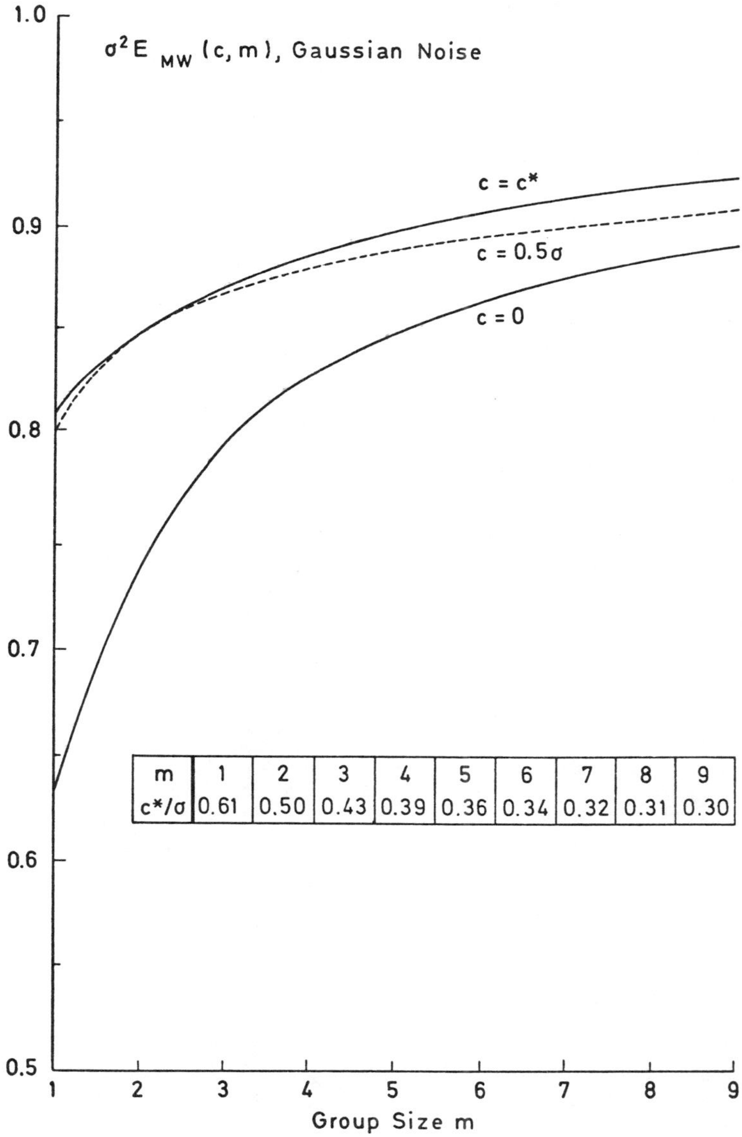

Figure 12 Efficacy of the conditional "mixed" Wilcoxon detector, Gaussian noise, variance σ^2.

Chapter 5

SMALL SAMPLE RESULTS

P. Papantoni-Kazakos

Department of Electrical Engineering
Rice University
Houston, Texas

I. INTRODUCTION

A large portion of the work done on nonparametric decision schemes is concerned with asymptotic results. Specifically, as Pyke points out [16, p. 21], it is empirically verified that 64% of all papers is nonparametric inference are concerned with such results. The emphasis on asymptotic studies is not necessarily an indication of the nature of the real problems. Indeed, the complexity involved in the small sample studies of nonparametric tests has forced the investigators to solve most of their problems in the paradise of infinity and to determine empirically or theoretically the ease (or difficulty) with which this paradise is achieved. However, infinity cannot be achieved always. This fact has been realized by several researchers [12-28] and empirical as well as analytical results have been obtained. Indeed, the necessity of more direct communication between the real world and the researcher has become more evident the last few years and the emphasis on realistic problems has increased.

One area that has experienced tremendous growth the last few years is the applications of statistical methods to clinical trials. In such trials the specification of the sample size for planning purposes before the experiment starts is a common problem. Often known nonparametric tests and power tables such as the ones included in this chapter can be used where the number of patients that will participate in the experiment is, in general, small.

Besides the clinical trials, the small sample nonparametric tests are applied to the study of communication, radar, or even social systems whenever the data used are small in number. These applications are known to the average reader of this book. Time or equipment limitations impose the necessity of the small sample consideration.

All nonparametric tests are based on the vector that is the result of ranking the available data and either taking their sign or indicating in which of two different groups they belong. This vector is called rank vector, it is denoted Z in this text and it has components that take each two possible different values.

In section II of this chapter, an outline of methods for computing the probability $P(Z)$ of the vector Z is given. In section III, powers and efficiencies of several known one-sample and two-sample rank tests are discussed. For both sections II and III probability, power, and efficiency tables are included. In section IV, a short description of sequential two-sample tests and some selected procedures are offered. Finally, in section V some ideas for future research are expressed.

II. THE DISTRIBUTION OF THE RANK VECTOR

A successful small sample behavioral analysis of rank tests must be preceded by a thorough study of the distribution of the rank vector Z. Indeed, as it will be shown in section III, all the interesting statistical parameters of these tests depend on the distribution $P(Z)$ of the vector Z.

The computational effort toward the calculation of these parameters can be reduced considerably if the proper $P(Z)$ representation is used. Also, due to efficient $P(Z)$ analysis, analytical evaluation of the rank tests can be achieved in certain specific cases.

The importance of the distribution $P(Z)$ was appreciated at the early development stages of the rank test. In 1951, Hoeffding first offered a simple formula for $P(Z)$ [8]. However, the difficulty in evaluating this formula makes it hard to use except in special cases. This difficulty has been partially overcome since by other means. Several investigators [5,9-11,13,15,25-27] studied empirically the power of one and two-sample rank tests and obtained, in some cases, curves. These results will be the subject of a following section.

In 1957, Tsao [25] presented a polynomial approximation of the distribution P(Z), while later Klotz [10] used a recursive formula to compute the same distribution in the presence of a small number of independent, normal data. Klotz studied also the behavior of P(Z) in the presence of high signal-to-noise ratio and data as described above. Another method was used for the same study by Davisson and me [15] and it was used to calculate the efficiencies of the optimum rank and the Wilcoxon nonparametric tests.

In this section, the representation methods of P(Z) that are considered more important are presented. The cases of one-sample and two-sample tests are examined separately and the data are considered independently.

A. One-Sample Tests: Arbitrary Hypothesis

The one-sample tests have been created to test the origins of a certain population of data when the alternatives are two. The data are either the outcomes of repeated trials from a general family of one-dimensional densities denoted as hypothesis H_o or they are the outcomes from a density family denoted as H_1. In all known studies on the rank distribution P(Z), the data have been taken independent from each other and hypothesis H_1 is a shifted version of hypothesis H_o. Then, each of the Z components takes two possible values depending on the sign of the observation it represents. In addition, the Z components are also independent from each other in this case. Under the assumption that the data are outcomes from a well-defined probability density f(x), a certain value Z_o of the vector Z represents the signs of them after they are ranked. If N is the number of the data, there are $N!$ possibilities of getting exactly the same data in different orders. Denote $P_o(Z_o)$ the probability of having the data ordered absolutely as received and with signs represented by Z_o. Then, the probability $P(Z_o)$ of obtaining a rank vector Z_o is given by

$$P(Z_o) = N! P_o(Z_o) \tag{1}$$

and the probability $P_o(Z_o)$ by

$$P_o(Z_o) = \underset{0 \le x_1 \le \cdots \le x_N \le \infty}{\int \cdots \int} \prod_{i=1}^{N} f(z_{oi} x_i) dx_i \tag{2}$$

where z_{oi} is the ith component of the vector Z_o. The expression (2) has been used as the basis for analytic studies of the distribution $P(Z_o)$. To achieve computationally convenient formulas for this distribution, a recursive method was developed [10].

Define

Z_N : an N-dimensional rank vector

Z_{oN} : a specific value of it

$A_{Z_{oN}}(u) = P\{Z_N = Z_{oN}$ and all data absolutely less than or equal to $u, u \ge 0\}$

and assume that the ith Z component is equal to one if the corresponding datum is positive and equal to zero otherwise.

Then, $A_{Z_{oN}}(\infty) = P(Z = Z_{oN})$ as defined by (1) and for data still being the outcomes from a probability density $f(x)$, it is also true that

$$\begin{aligned} A_{(1)}(u) &= P\{\text{the outcome of one trial from } f(x) \text{ is positive and absolutely not bigger than } u\} \\ &= F(u) - F(0) \end{aligned} \tag{3}$$

where

$$F(x) = \int_{-\infty}^{x} f(u)du \tag{4}$$

Similarly,

$$A_{(0)}(u) = F(0) - F(-u) \tag{5}$$

Let us now denote by $(Z_{oN,1})$ the vector that has the same components as the vector Z_{oN} up to the Nth one and the $(N+1)^{th}$ component equal to one. Then,

$$A_{(Z_{on,1})}(u) = P\{Z_{N+1} = (Z_{oN,1}) \text{ and all data absolutely not exceeding the value } u, u \geq 0\}$$

$$= \sum_{i=1}^{N+1} P\{Z_{N+1} = (Z_{oN,1}) \text{ , all data absolutely not exceeding the value } u \text{ , and the last taken datum to be the } i^{th} \text{ smallest in absolute value}\}$$

$$= (N+1)P\{Z_{N+1} = (Z_{oN,1}) \text{ , all data not exceeding absolutely } u \text{ , and the last datum to have also the absolutely highest value}\}$$

$$= (N+1)P\{Z_N = Z_{oN} \text{ , max (among } N \text{ data} \leq (N+1)^{th} \text{ datum} \leq u\}$$

$$= (N+1) \int_{0}^{u} P\{Z_N = Z_{oN} \text{ , all data absolutely not exceeding } x/\text{last datum equal to } x\} \cdot f(x)dx$$

$$= (N+1) \int_{0}^{u} P\{Z_N = Z_{oN} \text{ , all data absolutely not exceeding } x\} \cdot f(x)dx \tag{6}$$

Finally, we obtain from (6), substituting $A_{Z_{oN}}(u)$ in it

$$A_{(Z_{oN,1})}(u) = (N+1)\int_o^u A_{Z_{oN}}(x)f(x)dx \tag{7}$$

and similarly

$$A_{(Z_{oN,o})}(u) = \int_o^u A_{Z_{oN}}(x)f(-x)dx \tag{8}$$

The recursive nature of the last two formulas permits calculation of the P(Z) distribution with little computational effort. Some results on the power of the one-sample tests that are based on the recursive expressions (7) and (8) will be discussed in the following section. We must point out here that the recursive method described here is computationally superior to the polynomial approximation one that was proposed by Tsao [25]. We should also bring to the attention of the reader the fact that the recursive method is not the only one that is computationally good. Indeed, a composite midpoint quadrature approximation of $P_o(Z_o)$ in (2) was used by Milton in his effort to establish rank order probability and test power tables [13]. The method is structured in a way that takes under consideration the idiomorphism of the region of integration in (2).

As an example, let us consider the two dimensional integral

$$\iint\limits_{0 \le x_1 \le x_2 \le \infty} f(z_{o1}x_1)f(z_{o2}x_2)dx_1dx_2 \tag{9}$$

This integral is a special form of a more general one given by

$$I_2 = \iint\limits_{c \le x_1 \le x_2 \le d} g_1(x_1)g_2(x_2)dx_1dx_2 \tag{10}$$

where $g_1(\cdot)$ and $g_2(\cdot)$ are two different probability densities. The region of integration in (10) is indicated by the shaded area in Figure 1. If the interval [c,d] is divided into three equal subintervals of length ℓ on each axis, denote g_{1i}, g_{2i} the values of the densities g_1,g_2 correspondingly, at the midpoint of the ith interval. If the integrant at each midpoint is weighted by the area of the subregion of integration (rectangle or triangle) that includes the midpoint and the values on the diagonal boundary of the shaded region are weighted by $\frac{1}{2}$, the following approximation is obtained for I_2 .

$$I_2 = \ell^2 \left(\frac{g_{11} \cdot g_{21} + g_{12} \cdot g_{22} + g_{13} \cdot g_{23}}{2} + g_{11}g_{22} + g_{11}g_{23} + g_{12}g_{23} \right) = \ell^2 \left(\sum_{i=1}^{M} \frac{g_{1i}g_{2i}}{2!} + \sum_{1 \le i < j \le M} g_{1i}g_{2j} \right) \quad (11)$$

where the last part of expression (11) has been generalized to the case that the interval [c,d] is divided into M equal subintervals of length ℓ , on each axis.

Generalizing the midpoint procedure, as described above, to the k-dimensional integral I_k that is given by the expression

$$I_k = \int \cdots \int \prod_{i=1}^{k} g_i(x_i)dx_i \quad (12)$$

$$c \le x_1 \le x_2 \le \cdots \le x_k \le d$$

one obtains the following midpoint approximation for I_k

$$I_k = \ell^k \sum_{\alpha} \prod_{j=1}^{k} \frac{g_{ji}(\alpha)}{r(\alpha)} \quad (13)$$

where α is some ordering $1 \le i_1 \le i_2 \le \cdots \le i_k \le M$ with the relationship between two consecutive i_j's being either strict inequality or equality. All possible combinations of such strict inequalities and equalities give the space on which α takes values and $i(\alpha)$ is the value i in the jth place of the specific ordering α . $r(\alpha)$ is equal to the factorial of the number of unequal i_j's in the ordering α . A computing algorithm that applied on expression (13) economizes the number of arithmetic operations was developed by Savage and Milton [13].

This algorithm is discussed in more detail in subsection E of this section.

B. One-Sample Tests: Analytical Results

As we pointed out in the beginning of this section, the form of the region of integration in the expression of the distribution $P(Z)$ is such that analytical calculation of it is impossible, in general. However, there are some specific cases that allow such analytical handling. As it was pointed out by Savage [13, Foreword], most of these cases are of limited practical interest. For example, if the data are N in number and form an exchangeable process with some density function, each of the probabilities $P(Z_o)$ is equal to $1/N!$. Another case in which the probabilities $P(Z)$ can be computed is the case of independent data each distributed according to the distribution function F^{θ_i}, where F is continuous [18]. The same is also true if the data are uniformly distributed over intervals of different lengths. But the cases mentioned above are mostly interesting homework problems. Their physical significance is very limited. The interesting problem of Gaussian data cannot be studied analytically in general. However, it was found [9,15] that this is possible for the case of high signal-to-noise ratio. This case was first studied by Klotz [9] and later by Davisson and myself [15]. The results from the last reference will be presented here.

Let us suppose that the data available to us are independent, identically distributed and normal. Consider the case that their common mean is equal to λ, where λ is a large positive number and (without loss of generality) their variance is equal to 1. Furthermore, let us choose, for convenience, the values of Z components equal to ± 1, according to the sign of the corresponding datum. The probability

$P(Z_o)$ in (1) can be expressed, then, in the following way

$$P(Z_o) = N! \int \cdots \int \prod_{i=0}^{N} \varphi(x_i - z_{oi}\lambda)dx_i \tag{14}$$

$$0 \leq x_1 \leq \ldots \leq x_N \leq \infty$$

where

$$\varphi(x) = (2\pi)^{-\frac{1}{2}}\exp\{-\tfrac{1}{2}x^2\}$$

Because of the high value of λ, the integral in (14) can be broken into a product of integrals with lower dimensionality if the tail values are ignored.

The critical measure on which this breaking depends is the vector of the x_i values that maximize the integrand in (14). This vector will be denoted X_{max} and its components $x_{i,max}$, where $1 \leq i \leq N$. If the integral in (14) is separated into clusters corresponding to consecutive, equally valued $x_{i,max}$ components, the probability $P(Z_o)$ can be expressed as

$$P(Z_o) = N! \prod_{i=1}^{n} P_i \tag{15}$$

where n is the number of the groups of consecutive equally values $x_{i,max}$ components and the P_i's are partial integrals given by

$$P_i = \int \cdots \int \prod_{j=1}^{m_i} \varphi(x_{ij} - z_{o,M_{i-1}+j} \cdot \lambda)dx_{ij} \tag{16}$$

$$0 \leq x_{i1} \leq \ldots \leq x_{im_i} \leq \infty$$

In (16), the x_{ij}'s, $j = 1,\ldots,m_i$ correspond to the ith cluster of consecutively equal $x_{i,max}$'s, m_i is the number of members in that cluster and

$$M_i = \sum_{j=1}^{i} m_j$$

The integrals P_i can actually be calculated, but for them to be determined, first the vector X_{max} must be determined. An interesting dynamic programming procedure that leads to X_{max} is presented in appendix A. After X_{max} is determined, the integrals P_i are well defined and available for calculation.

If the integers K_i and μ_i that were defined in appendix A are used, the exponent in the integrand of (16) can be expressed as

$$\sum_{j=1}^{m_i} \left(x_{ij} - z_{o,M_{i-1}} + j \cdot \lambda\right)^2 = \sum_{j=1}^{m_i} \left[\left(x_{ij} - \frac{k_i - \mu_i}{k_i + \mu_i} \cdot \lambda\right) + \left(\frac{k_i - \mu_i}{k_i + \mu_i} \lambda - z_{o,M_{i-1}+j} \cdot \lambda\right)\right]^2 \quad (17)$$

where (as concluded from appendix A)

$$\begin{aligned} k_i + \mu_i &= m_i \\ k_i - \mu_i &= \sum_{k=1}^{m_i} z_{o,M_{i-1}+k} \end{aligned} \quad (18)$$

By substituting (18) in (17), we can easily obtain

$$\sum_{j=1}^{m_i} \left(x_{ij} - z_{o,M_{i-1}+j} \cdot \lambda\right)^2 = \sum_{j=1}^{m_i} \left[\left(x_{ij} - \frac{\sum_{k=1}^{m_i} z_{o,M_{i-1}+k}}{m_i} \lambda\right) + \left(\frac{\sum_{k=1}^{m_i} z_{o,M_{i-1}+k}}{m_i} - z_{o,M_{i-1}+j} \cdot \lambda\right)\right]^2 \quad (19)$$

The final form of the expression in (19) will depend on the mutual relationship between the k_i's and the μ_i's. The several possibilities are explored.

1. $k_i > \mu_i$ CASE

Then, the expression in (19) becomes

$$\sum_{j=1}^{m_i}\left(x_{ij} - z_{o,M_{i-1}+j}\cdot\lambda\right)^2 = \sum_{j=1}^{m_i}\left(x_{ij} - \frac{\sum_{k=1}^{m_i} z_{o,M_{i-1}+k}}{m_i}\lambda\right)^2$$

$$+ 2\lambda\sum_{j=1}^{m_i} x_{ij}\left(\frac{\sum_{k=1}^{m_i} z_{o,M_{i-1}+k}}{m_i} - z_{o,M_{i-1}+j}\right) + \lambda^2\sum_{j=1}^{m_i}\left(1 - \frac{\left[\sum_{k=1}^{m_i} z_{o,M_{i-1}+k}\right]^2}{m_i^2}\right) \tag{20}$$

If the new parameter

$$\nu_i = \frac{\sum_{k=1}^{m_i} z_{o,M_{i-1}+k}}{m_i}, \quad -1 \le \nu_i \le 1$$

is defined and it is substituted in (20), the following expression is obtained

$$\sum_{j=1}^{m_i}(x_{ij} - z_{o,M_{i-1}+j}\cdot\lambda)^2 = \sum_{j=1}^{m_i}(x_{ij} - \nu_i\lambda)^2 + 2\lambda\sum_{j=1}^{m_i} x_{ij}(\nu_i - z_{o,M_{i-1}+j})$$
$$+ \lambda^2(1 - \nu_i^2)m_i \tag{21}$$

In (21), no operations with respect to j are included and since $k_i > \mu_i$, ν_i is positive. Also, since λ is large, so is $\nu_i\lambda$, and if tail values are ignored and (21) is substituted in (16), the integration in the last expression can be extended to minus infinity. If, in addition, the new constants

$$\begin{aligned} b_{ij} &= \lambda\left[z_{o,M_{i-1}+j} + a_j\right] \\ a_i &= \sqrt{1 - \nu_i^2} \end{aligned} \tag{22}$$

are introduced and change of variables is applied, the integral P_i becomes

$$P_i = \exp - \frac{m_i(1 - \nu_i^2)}{2}\lambda^2 \cdot \int\cdots\int \prod_{j=1}^{m_i}\left[\varphi(x_{ij})\exp(-\lambda x_{ij}c_j)\right]dx_{ij}$$
$$-\infty \le x_{i1} \le \cdots \le x_{im_i} \tag{23}$$

where

$$c_i = \begin{cases} a_i & , x_{ij} \geq 0 \\ -a_i & , x_{ij} < 0 \end{cases}$$

The integral in (23) represents the probability of receiving m_i data in a way that the ith in time is also the ith in absolute magnitude. The probability of receiving the same data in an arbitrary order is given by

$$A^{m_i} \prod_{i=1}^{m_i} \left[\int_{-\infty}^{\infty} \varphi(x)\exp(-\lambda x c_i)dx \right] \tag{24}$$

where A is the normalizing constant.

The integral in (23) if multiplied by $A^{m_i} m_i!$ should equal to the expression in (24) then. Due to that, by direct substitution, we obtain

$$P_i = \exp\left(- \frac{m_i(1 - \nu_i^2)}{2} \lambda^2 \right) \cdot \frac{1}{m_i!} \prod_{j=1}^{m_i} \left[\int_{-\infty}^{\infty} \varphi(x)\exp(-\lambda x c_i)dx \right] \tag{25}$$

The product $\lambda x c_i$ is always positive in the case examined. Therefore, the product $\varphi(x)(-\lambda x c_i)$ is bounded from above by $\varphi(x)$ and an upper bound to P_i is consequently given by

$$\frac{1}{m_i!} \exp \left(- \frac{m_i(1 - \nu_i^2)}{2} \lambda^2 \right) \tag{26}$$

Also, since

$$\frac{1}{m_i!} \prod_{j=1}^{m_i} \left[\exp(-\lambda E\{xc_i\}) \right] - \frac{1}{m_i!} \prod_{j=1}^{m_i} \left[E\{\exp(-\lambda x c_i)\} \right]$$

due to the convexity of the exponential function, it is obvious that a lower bound to P_i is given by

$$\frac{1}{m_i!} \exp \left(- \frac{m_i(1 - \nu_i^2)}{2} \lambda^2 - \lambda \sum_{j=1}^{m_i} E\{xc_i\} \right) \tag{27}$$

Since $\nu_i \neq 1$, the term

$$\exp\left(-\frac{m_i(1-\nu_i)}{2}\lambda^2\right)$$

is the predominant one in the bound in (27) for $\lambda \to \infty$. Then, the bounds (26) and (27) approach and due to that we obtain

$$\lim_{\lambda\to\infty} P_i = \frac{1}{m_i!}\exp\left(-\frac{m_i(1-\nu_i^2)}{2}\lambda^2\right) \tag{28}$$

2. $k_1 = 0$ CASE

As shown in appendix A, the only group of ± 1 components that has a number of +1's smaller than the number of -1's is the first one. Then, in the case of $k_1 = 0$, all the components of the first group are -1's and the multiple integral P_1 is given by

$$P_1 = \int\cdots\int \prod_{j=1}^{m_i} \varphi(x_j + \lambda)dx_j \tag{29}$$
$$0 \le x_1 \le \ldots \le x_{m_1} \le \infty$$

Following procedure similar to the one in 1, we obtain from (29)

$$P_1 = \frac{1}{m_1!}\left[\int_0^\infty \varphi(x+\lambda)dx\right]^{m_1} \tag{30}$$

Since

$$\lim_{\lambda\to\infty}\left[\int_0^\infty \varphi(x+\lambda)dx\right] = \lim_{\lambda\to\infty}\left[\int_{-\infty}^{-\lambda}\varphi(x)dx\right] = \frac{\varphi(-\lambda)}{\lambda} \tag{31}$$

and by direct substitution, the following expression is obtained from (29)

$$P_1 = \frac{1}{m_1!\lambda^{m_1}}\exp\left(-m_1\frac{\lambda^2}{2}\right) \tag{32}$$

3. $k_1 \le \mu_1$ CASE

This case represents a Z vector with components forming only one group, as shown in appendix A. Then, an expression for P_1 is obtained that is very similar to the one in (23). The difference is that in the present case, ν_1 is negative. The factor $\varphi(x - \nu_1\lambda)$ is the predominant one in the product $\varphi(x - \nu_1\lambda)\exp(-\lambda x[\nu_1 - z_{oj}])$ for large λ and P_1 can be expressed then as follows

$$P_1 \simeq \exp\left(- \frac{m_1(1 - \nu_1^2)}{2}\lambda^2\right) \underset{0 \le x_1 \le \dots \le x_{m_1} \le \infty}{\int \dots \int} \prod_{j=1}^{m_1} \varphi(x_j - \nu_1\lambda)dx_j \tag{33}$$

If the concept of ordered and unordered probabilities is again used here, the following expression is obtained for P_1

$$P_1 \simeq \frac{1}{m_1!} \exp\left(- \frac{m_1(1 - \nu_1^2)}{2}\lambda^2\right)\left[\int_o^\infty \varphi(x - \nu_1\lambda)dx\right]^{m_1} \tag{34}$$

Methods used in case 2 lead finally to

$$\lim_{\lambda\to\infty} P_1 = \frac{1}{(\nu_1\lambda)^{m_1}} \exp(- \frac{m_1}{2}\lambda^2) \tag{35}$$

Here, $m_1 = N$ (number of data).

Expressions (28), (32), and (35) give the three possible analytic revisions of the multiple integral P_i defined in (16) when the data are independent and normal with very high mean λ and the components of the rank vector are equal to $\pm$ 1.

From expression (15) and the formulas giving the partial integrals P_i, the expression for the probability $P(Z_o)$ is easily obtained. Indeed, a uniform formula covering all three cases (1, 2, and 3) is easily developed.

$$\lim_{\lambda\to\infty} P(Z_o) = K(Z_o) \cdot \lambda^{-s_1} \exp\left(- \frac{\lambda^2}{2} \sum_{i=1}^{n} m_i\rho_i^2\right) \tag{36}$$

where

$$s_1 = \begin{cases} m_1 & , \mu_1 \geq k_1 \\ 0 & , \mu_1 < k_1 \end{cases}$$

$$\rho_1 = \begin{cases} \nu_i & , \nu_k > 0 \\ 0 & , \nu_i \leq 0 \end{cases} \tag{37}$$

and the μ's and k's are explained in appendix A. Also, $K(Z_o)$ is a function of the given rank vector Z_o only.

Expression (36) is the only analytical closed expression of the probability $P(Z_o)$ for finite number of samples and Gaussian data.

C. Two-Sample Tests: Arbitrary Hypotheses

The two-sample tests have been created for testing the equality of the distributions of two different groups of samples. All the samples are ranked and a rank vector Z is formed whose i^{th} component takes one of two values (usually 0 or 1) depending on if the ith smallest (not absolutely) datum from sample group number one or sample group number two. In the known studies of the rank distribution $P(Z)$, when Z has finite dimensionality, the samples have been assumed independent from each other. There the Z components are independent also and distributed according to the probability densities $f(x)$ and $g(x)$ that represent each of the two sample groups.

Let us suppose that the two-sample test is to be tested as to its ability to discriminate between two well-defined densities $f(x)$ and $g(x)$. Let m independent samples be taken from density $f(x)$ and n also independent samples from density $g(x)$. If all these $m+n$ data are combined and ranked, an $m+n$-dimensional vector of zeros and ones will be formed as explained in the beginning of this subsection.

Call Z_o a particular value of this vector and $P_o(Z_o)$ the probability of the data being received in the order represented by Z_o. Then

$$P_o(Z_o) = \int \ldots \int \prod_{i=1}^{m+n} f^{1-z_{oi}}(x_i) g^{z_{oi}}(x_i) dx_i \qquad (38)$$

$$-\infty \le x_1 \le \ldots \le x_{m+n} \le \infty$$

where z_{oi} the ith component of Z_o.

The probability $P(Z_o)$ of obtaining the rank vector Z_o will be produced from $P_o(Z_o)$ if the last is multiplied by the possible ways the data can be received and still produce Z_o when ordered. These possibilities are $m!n!$ in number and therefore

$$P(Z_o) = m!n! \, P_o(Z_o) \qquad (39)$$

A recursive method that leads to numerical computation of $P(Z_o)$ and is similar to the one presented in subsection A of this chapter has been developed in this case also.

We define again

Z_N : an N-dimensional rank vector where $N = m+n$ here

Z_{oN} : a specific value of Z_N

$A_{Z_{oN}}(u) = P\{Z_N = Z_{oN}$ and all data of value that is bounded from above by $u\}$ (40)

Then $A_{Z_{oN}}(\infty) = P(Z_{oN})$ (as defined in (39)).

Let us now consider the vector $Z_{o,N+1} = (Z_{oN}, 1)$ that coincides with Z_{oN} in all first N components and has one more component (last one) that is equal to one. Then the following relationships are obtained

$$A_{(Z_{oN,1})}(u) = P\{Z_{N+1} = (Z_{oN,1}) \text{ and all samples with values that do not exceed } u\}$$

$$= \sum_{i=1}^{m+1} P\{Z_{N+1} = (Z_{oN,1}), \text{ all samples of values that do not exceed } u \text{ , the last datum taken is the } i^{th} \text{ smallest in value}\}$$

$$= (m+1)P\{Z_{N+1} = (Z_{oN,1}), \text{ all values not exceeding } u, \text{ the last datum taken is } (m+1)^{th} \text{ smallest in value}\}$$

$$= (m+1)P\{Z_N = Z_{oN}, \max \text{ (among m+n data)} \leq (m+n+1)^{th} \text{ datum} \leq u\}$$

$$= (m+1)\int_{-\infty}^{u} P\{Z_N = Z_{oN}, \max \text{ (among m+n data)} \leq x/\text{last datum equal to } x\} \cdot f(x)dx \qquad (41)$$

If definition (40) is considered, the following recursive expression is obtained from (41)

$$A_{(Z_{oN,1})}(u) = (m+1)\int_{-\infty}^{u} A_{(Z_{oN})}(x)f(x)dx \qquad (42)$$

where it has been assumed that the value 1 is assigned to a certain component of the rank vector Z , if the corresponding datum comes from the probability density $f(x)$. In this case it is similarly found that:

$$A_{(Z_{oN,o})}(u) = (m+1)\int_{-\infty}^{u} A_{(Z_{oN})}(x)g(x)dx \qquad (43)$$

As in the one-sample tests, the recursive formulas (42) and (43) are not the only ones representing computationally good approaches to the calculation of the rank probability $P(Z_o)$. The computational method that was used by Milton [13] can be applied here identically as in the one-sample case, if $g_i(x_i)$ in (12) is representing the product

$$f^{1-z_{oi}}(x_i)g^{z_{oi}}(x_i)$$

in (38). Results from the application of this method to normal data will be presented in subsection E of this section.

D. Two-Sample Tests: Normal Data

An interesting application of two-sample testing consists of Gaussian statistics and location checking.

In that, the equality of the means of two normal probability densities with common variance is tested from two sample groups, one from each density. The testing of signal equality when embedded in Gaussian white noise with known statistical parameters falles in this category. In this case, the product

$$f^{1-z_{oi}}(x_i)g^{z_{oi}}(x_i)$$

in (38) is given by

$$f^{1-z_{oi}}(x_i)g^{z_{oi}}(x_i) = \varphi^{1-z_{oi}}(\frac{1}{\sigma}[x_i - \mu_1])\varphi^{z_{oi}}(\frac{1}{\sigma}[x_i - \mu_2]) \quad (44)$$

where $\varphi(\cdot)$ the normal density, σ^2 the common variance, and μ_1, μ_2 the means whose equality is to be tested. Directly from (44) and considering the exponential nature of $\varphi(\cdot)$, we obtain

$$f^{1-z_{oi}}(x_i)g^{z_{oi}}(x_i) = \varphi(t_i - z_{oi}d) \quad (45)$$

where

$$t_i = \frac{x_i - \mu_1}{\sigma} \quad (46)$$

$$d = \frac{\mu_2 - \mu_1}{\sigma} \quad (47)$$

The ranking of the t_i's in (46) follows exactly the ranking of the x_i's of course. Therefore, the probability $P(Z_o)$ in (39) takes the following form here

$$P(Z_o) = m!n!\int\cdots\int \prod_{i-1}^{N+m+n} \varphi(t_i - z_{oi}d)dt_i \qquad -\infty \le t_1 \le \cdots \le t_{m+n} \le \infty \quad (48)$$

The only difference between the expression in (48) and the one in (14) is the multiplying constants $N!$ and $m!n!$. For distance d assumed large (far apart densities) the same analysis of $P(Z_o)$ that was presented in the previous subsection is also valid here. If the distance d can be assumed very small a priori, applying a series expansion introduced by Terry [25], an interesting approximation of the $P(Z_o)$ in (48) is obtained. Obviously, the same approximation can be used for the one-sample case where (14) is the proper $P(Z_o)$ expression. Let us express $P(Z_o)$ as a function of d also. That is, let us define

$$P(Z_o,d) = m!n! \int\cdots\int \prod_{i=1} \varphi(t_i - z_{oi}d)dt_i \tag{49}$$
$$-\infty \le t_1 \le \cdots \le t_{m+n} \le \infty$$

and

$$c_k(Z_o) = \left.\frac{\theta^k P(Z_o,d)}{\theta d^k}\right|_{d=0} \tag{50}$$

the k-order continuous derivative of $P(Z_o,d)$ with respect to d and at the point $d = 0$.

Then

$$P(Z_o,d) = \sum_{k=0}^{\infty} c_k(Z_o) \frac{d^k}{k!} \tag{51}$$

if the convergence of the series is not considered.

If the integrand in (49) is expanded in the powers of d and it is then substituted in (50), the coefficients will be obtained as functions of d and derivatives of $\varphi(\cdot)$. Indeed, the series expansion of $\varphi(t_i - z_{oi}d)$ gives

$$\prod_{i=1}^{m+n} \varphi(t_i - z_{oi}d) = \sum_{k=0}^{\infty} \Sigma^{(k)} \frac{k!}{\prod_{i=1}^{m+n} k_i!} \prod_{i=1}^{m+n} z_{oi}^{k_i} \varphi^{(k_i)}(t_i)\frac{(-d)^k}{k!} \tag{52}$$

where $\varphi^{(k_i)}(t_i)$ is the k_i^{th} derivative of $\varphi(t_i)$ at point t_i and $\Sigma^{(k)}$ represents the summation over all $k_1, k_2, \ldots, k_{m+n}$ such that $k_i \geq 0$, $i = 1,\ldots,m+n$ and

$$\sum_{i=1}^{m+n} k_i = k$$

From (50) and (52) jointly, the following expression is obtained for $c_k(Z_o)$

$$c_k(Z_o) = m!n! \int \cdots \int \Sigma^{(k)} \frac{k!}{\prod_{i=1}^{m+n} k_i!} \prod_{i=1}^{m+n} z_{oi}^{k_i} (-1)^{k_i} \varphi^{(k_i)}(t_i) dt_i \tag{53}$$

$$-\infty \leq t_1 \leq \ldots \leq t_{m+n}$$

Considering the equation

$$\varphi^{(k)}(x) = (-1)^k H_k(x)\, \varphi(x) \tag{54}$$

where $H_k(x)$ is the kth order Hermite polynomial, we obtain the following new form for $c_k(Z_o)$ in (53)

$$C_k(Z_o) = m!n! \int \ldots \int \Sigma^{(k)} \frac{k!}{\prod_{i=1}^{m+n} k_i!} \prod_{i=1}^{m+n} z_{oi}^{k_i} H_{k_i}(t_i)\, \varphi(t_i) dt_i$$

$$-\infty \leq t_1 \leq \ldots \leq t_{m+n} \leq \infty$$

$$= \frac{m!n!}{(m+n)!} E\{\Sigma^{(k)} \frac{k!}{\prod_{i=1}^{m+n} k_i!} \prod_{i=1}^{m+n} z_{oi}^{k_i} H_{k_i}(t_{(m+n)r_i})\} \tag{55}$$

In (55), $t_{(m+n)1} \leq t_{(m+n)2} \leq \cdots \leq t_{(m+n)(m+n)}$ are the order statistics in a normal random sample, and

$$(m+n)! \prod_{i=1}^{m+n} \varphi(x_i)$$

$x_1 \le x_2 \le \dots \le x_{m+n}$ their joint frequency function. From (55) the first three $C_k(Z_o)$ coefficients can be found in particular. They are given by

$$C_o(Z_o) = \frac{m!n!}{(m+n)!}$$

$$C_1(Z_o) = \frac{m!n!}{(m+n)!} \sum_{i=1}^{m+n} z_{oi}\, E\{t_{(m+n)r_i}\} \tag{56}$$

$$C_2(Z_o) = \frac{m!n!}{(m+n)!} \left[E\left\{ \sum_{i=1}^{m+n} z_{oi}\, t_{(m+n)r_i} \right\}^2 - \sum_{i=1}^{m+n} z_{oi}^2 \right]$$

If the z_{oi}'s take the values 0 and +1 , the sum

$$\sum_{i=1}^{m+n} z_{oi}^2$$

is, of course, equal to

$$\sum_{i=1}^{m+n} z_{oi}$$

If the distance d is so small that the powers of it that are higher than two are negligible, the probability $P(Z_o,d)$ in (49) can be approximated by the first three terms in (51). That is

$$P(Z_o,d) \simeq C_o(Z_o) + C_1(Z_o) \cdot d + \tfrac{1}{2}C_2(Z_o)d^2 \tag{57}$$

then, where $C_o(Z_o)$, $C_1(Z_o)$ and $C_2(Z_o)$ are given by (56). For d small enough, the coefficient $C_1(Z_o)$ is the one that characterizes the rank probability $P(Z_o,d)$. Also, the locally (small d) most powerful rank test corresponds to the Z_o vectors that maximize $C_1(Z_o)$. Therefore, the $C_1(Z_o)$ statistics are important to be investigated. Analytically, these statistics have been found only in the asymptotic case of m and n been very large. For small sample sizes, $C_1(Z_o)$ has been investigated numerically by Terry [25] and Milton [13] and tables giving its values for different m and n sample sizes have been formed. More detailed discussion on that is presented in subsection E of section III.

E. Numerical Results

The method that was first used for calculation of the probability $P(Z_o)$, is the recursive one first expressed by equation (8) of this section and then by equations (42) and (43). Klotz [10] used the last two recursive formulas in Gaussian two-sample testing and shift alternatives. In particular, the densities $g(x) = \varphi(x)$ and $f(x) = \varphi(x - \mu)$ were used as alternatives, where $\varphi(\cdot)$ is the normal density and μ a location parameter. Let M be the number of steps used for a single infinite integration and let N_{max} represent the dimensionality of the lengthier Z_o vector whose probability will be calculated. Then, the storage space required before any reduction is attempted is given by the number

$$M \times \left[2^{N_{max}+1} - 2\right]$$

[10, p. 662]. By efficient ordering in the computing of

$$A_{(Z_{oN})}(u)$$

this space can be reduced to $M \cdot N_{max}$. An example of this ordering is shown below for the case of $N_{max} = 3$. This is shown in Fig. 2. There are still

$$2^{N_{max}+1} - 2$$

locations that are needed for the answers

$$A_{(Z_o)}(\infty)$$

In Fig. 2, it is obvious, for example, that once the integrals for (1,1,1) have been computed and $A_{(1,1,1)}(\infty)$ is stored, they are no longer needed for recursion and the integrals for (1,1,0) can be built up using (43) and the values $A_{(1,1)}(u)$. Because of the error accumulation, Klotz points out that a 5th or 6th degree integration rule such as Weddle's works better than does Simpson's rule. Klotz also checked the accuracy of his results by summing the probabilities and comparing with one and by setting $\mu = 0$ and comparing the values with $1/\binom{N}{m}$. Another accuracy check was obtained by using the symmetry that was noted by Savage [18, p. 975]. Savage observed that if two vectors $Z_{1N} = (z_{11},\ldots,z_{1N})$ and $Z_{2N} = (z_{21},\ldots,z_{2N})$ are such that $z_{1,i} = 1 - z_{2,N-i+1}$, for every $i = 1,\ldots,N$, then $P(Z_{1N}) = P(Z_{2N})$. Klotz believed his results to be accurate to within one unit in the last significant figure. The vectors Z_N in his table, which are composed of 0's and 1's are treated as binary numbers and converted to octal numbers. This was done to save space. Additional saving in space is achieved if only one of the equiprobable vectors Z_{1N} and Z_{2N} is listed.

In a previous work [9], Klotz calculated the expected values of the absolute normal order statistics that appear in one-sample tests. The vector Z_o was assumed to have 0 and 1 components and the alternatives were one-sided shift ($\mu > 0$) . The recursive expression in (8) was used and and a table was formed. This table is obtained under the null hypothesis and Z vectors of dimensionality from 1 to 10 are considered. E_{N_i} are numbers equal to the expected value of the ith smallest order statistic from a sample of N absolute normal variables. As it will be shown in section III of this chapter, the expected values E_{N_i} are used in the study of the normal scores test.

Milton [13] is the last known investigator to have calculated numerically rank probabilities and powers of two-sample tests. As it was mentioned in subsection A of this section, he used a midpoint quadrature approximation to compute multiple integrals as the one in (12). The mid-point approximation he ended up with in these cases is given by (13) (explained at the point). Savage and Milton developed a computing algorithm 13, p. 10 for evaluating I_k in (13) that economizes the number of arithmetic operations. This algorithm works in the following way.

First define

$S_j = (s_{j1},\ldots,s_{j,M+1})$ such that $I_j = \ell^j s_{j,M+1}$ where I_j is defined in (13)

G_j = a vector with points the values of $g_j(\cdot)$ evaluated at the midpoints of the M equal subintervals of length $\ell = (d - c)/M$ of the interval $[c,d]$. So, g_{ji} (in (13)) $= g_j(x_i)$ where

$$x_i = c + \frac{(d - c)(2i - 1)}{M}$$

$\mathcal{A}$: operator that applied to a vector $X = (x_1,\ldots,x_N)$ gives a vector Y such that

$$\mathcal{A}X = Y = (0,x_1,\ldots,\sum_{i=1}^{p-1} x_i,\ldots, \sum_{i=1}^{N} x_i) = (y_1,y_2,\ldots,y_{N+1})$$

X*Y: inner product of two vectors with the same dimensionality

$V_{11} = G_1$

$S_1 = \mathcal{A}V_{11}$

Then follow the following procedure

1) For k = 1 (k as given by (13)), calculate V_{11} and S_1 .

2) For $k > 1$ and $j = 2,\ldots,k$, calculate

$$V_{ji} = V_{j-1,i}*G_j/(j+1=i),\ i < 1$$

$$S_j = \mathcal{A}(\sum_{i=1}^{j} V_{ji})$$

$$V_{jj} = S_{j-1}*G_j$$

Obviously, as $\ell \to 0$, $\ell^k s_{k,M+1} \to I_k$ in (13). On page 11 of his book, Milton [13] gives details on the calculations accomplished. He explains that for the evaluation of the probability $P(Z_o)$ in (48) the range of integration was truncated to (-8,8) and that the increment used was $\ell = .01$. Also, it was observed that for single precision evaluation of (48) with $k = m+n$, the vectors $V_{k1},\ldots,V_{k6}$ were sufficient for the approximation of I_k, where $k > 6$. To further improve the error in the algorithm, Milton used also the technique of "extrapolation to the limit" which he describes on pages 11, 12, and 13 of his book. In short, a series of approximations $A_{i,o}$ was produced for $P(Z_o)$ in (48) through the use of the algorithm explained above. Then a triangular extrapolation array was formed such that

$$\begin{array}{llll} A_{o,o} & & & \\ A_{1,o} & A_{1,1} & & \\ A_{2,o} & A_{2,1} & A_{2,2} & \\ A_{3,o} & A_{3,1} & A_{3,2} & A_{3,3} \end{array} \qquad A_{i,p+1} = \frac{2^{2p+2}A_{i,p} - A_{i-1,p}}{2^{2p+2}-1}, \quad \begin{array}{l} p = 0,1,\ldots,9 \\ i = p,p+1,\ldots,8 \end{array}$$

It was shown that the error in approximating (48) in this manner is described by the equation

$$\frac{A_{i,p} - A_{i-1,p}}{A_{i+1,p} - A_{i,p}} = 4^{p+1}$$

Milton continued his calculations until $|A_{i,p} - A_{i-1,p}| < 10^{-9}$. This method was used for m,n combination large enough. Specifically, the cases of $m = n = 7$ and $m = 12$, $n = 1$ were computed through this method. For smaller values of m and n Savage's back-recursive rule [20] was used. The checks on the accuracy of the computations that were applied are explained in pages 13 through 19 of Milton's book [13]. The results of the $P(Z_o)$ computations as given by (48) are given in table A of the same book.

The c_1 and c_2 in table A [13] correspond to $C_1(Z_o)$ and $C_2(Z_o)$ in (56) and w is the Wilcoxon two-sample expressions. Also, the Z vectors are ordered there according to the decreasing order of their c_1 values. Z^{+c} is the vector whose i^{th} component is equal to $1-z_{m+n+1-i}$ where z_j is the jth component of the vector Z. Studying correspondence between the tables in [10] and [13] is not straightforward since the ordering in them is according to different principles. However, for the shifting values 1.5, 2.0 and 3.0 that are included in both tables and if the ordering is decoded, a good correspondence is noticed. Higher accuracy is observed though in the tables in [13]. This was expected, since the computational algorithm and the accuracy checkings used there are more sophisticated.

We will conclude this section by commenting on a Hodges and Lehman [7] evaluation of the $P(Z_o)$ probability in (48) when the location parameter has a high value. If a given rank vector Z has its components grouped in clusters of equally valued consecutive ones, and if the 0 groups are indicated by r and +1 groups by s, a sequence of r_i's and s_i's is formed. r_i represents the number of zeros in the ith zero group and s_i the corresponding number in the ith+1 group, and the sequence

$$Z = (\underbrace{0\ldots0}_{r_o}\ \underbrace{1\ldots1}_{s_1}\ \underbrace{0\ldots0}_{r_1}\ .\ .\ .\ \underbrace{1\ldots1}_{s_o})$$

starts by r_o and ends by s_o always. A graph can be obtained if each 0 is represented by a horizontal and each +1 by a vertical unit segment. Such a graph is shown in Fig. 3. The lower convex hull represented by the dotted line is the interesting characteristic here. Hodges and Lehman proved that if Z is not of the for (0...0 1...1), then

$$\lim_{d\to\infty}[P(Z_o)^{d^{-2}}] = \exp\left(-\tfrac{1}{2}\sum_{i=1}^{k}\frac{a_ib_i}{a_i+b_i}\right) \tag{58}$$

where $a_i = A_i - A_{i-1}$ and $b_i = B_i - B_{i-1}$, $i = 1,\ldots,k$, (A_j, B_j) are the coordinates of the $(j+1)^{th}$ corner point of the convex hull and $k+1$ the number of these points. We must point out here that the formula in (58) represents a different method for evaluating analytically $P(Z_o)$ than the one described in subsection B of this section. Table 1 in [13] gives the values of the probability $P(Z)$ for two specific rank vectors $Z_o = (0\ldots0, 1\ldots1)$ and $Z_1 = (0\ldots0101\ldots1)$, different m,n sample sizes and values of distance d between 4 and 6. The computational algorithm described in this subsection was used and the results from expression (58) for the same vector. The maximum departures above and below the Hodges-Lehman value o occur for $(m,n) = (7,7)$ and $(1,1)$ for all values of d.

F. Conclusion

In this section, a rather complete description of the studies that have been done on the $P(Z)$ probability was offered. Analytical calculation of $P(Z)$ was possible only in isolated cases. Emphasis has been given by most investigators to the development of computationally efficient methods for calculating $P(Z)$ in the presence of small number of data. Tables of $P(Z)$ values have been formed, a short discussion of which is offered in this chapter.

III. SMALL SAMPLE POWER AND EFFICIENCY OF RANK TESTS

Before the known rank tests are studied in the presence of small samples, the t-test should be mentioned. The one-sample t-test has been used as a basis for evaluating rank tests [9] and if $x_1,\ldots,x_n$ are the available data, it consists of testing the expression

$$\frac{\frac{1}{\sqrt{n}}\sum_{i=1}^{n} x_i}{\frac{1}{n-1}\sum_{i=1}^{n}\left(x_i - \frac{1}{\sqrt{n}}\sum_{j=1}^{n} x_j\right)^2} \qquad (59)$$

against a threshold C , and deciding accordingly. The denominator in (59) converges to the variance and the numerator is asymptotically Gaussian by the central limit theorem. The test is, therefore, asymptotically nonparametric in the sense that the false alarm rate becomes fixed for any probability distribution with finite variance and independent samples, as $n \to \infty$.

The rank tests that developed later have nonparametric properties not only asymptotically, but also when the number of data used is small. However, because of the analytical difficulties involved in the small sample cases, the studies on rank tests were primarily asymptotic for many years. Terry [25] first developed tables of the $C_1(Z_o)$ test as presented in (56) for small m and n sample numbers and Gaussian independent data. Lehman [11] developed some distribution curves of the two-sample rank data almost simultaneously with Terry.

All the known completed studies on small sample rank tests have assumed independent normal data. This assumption combines practical interests and analytical convenience. Inside the framework of these assumptions, Teichroew [22] calculated the power for the two-sample $C_1(Z_o)$ test in (56) for several sample values m and n , while Tsao [25] calculated just probabilities of the rank vector using his polynomial approximation.

With all the emphasis in the two-sample tests, the study of the behavior of the one-sample ones in the presence of small samples was neglected. Klotz [9] was the first to develop power tables for the Wilcoxon and normal scores one-sample tests for Gaussian shift alternatives. Later [10], he also studied the two-sample normal scores test and presented distribution and approximation tables for it. Milton [13], using computationally efficient methods as shown in the previous section, managed to exhibit a rather complete selection of power tables for several known rank tests. Fianlly, the author together with Davisson [5,15], offered an analytical result for the efficiency of the optimal rank and Wilcoxon one-sample tests for normal for apart shift alternatives.

In this section a selection of results on the small-sample power and efficiency of rank tests is presented.

A. One-Sample Optimum Rank Test

The power and efficiency expressions for rank test are directly related to the probability $P(Z)$ of the rank-vector Z. The mathematical complexity involved in the analytical calculation of $P(Z)$ is the primary cause of the analytical results on rank tests being limited. As explained in subsection B of section II, a practically interesting analytically managable rank situation is the one of normal, independent data and far distance shift alternatives. Specifically, let H_o and H_1 be the two hypothesis representing normal, variance one (without loss in generality), mean $-\lambda$ versus $+\lambda$ one-dimensional densities. It is assumed that N independent data originating from one of the two densities are received and the objective is to decide on the true hypothesis using a rank one-sample test. In subsection B of section II, the analytical expression (36) for probability $P(Z)$ was given under the H_1 hypothesis for very large value of the mean λ. Let us observe here that the probability $P_{H_o}(Z_o)$ of observing the rank vector Z_o under the hypothesis that it originates from the density corresponding to hypothesis H_o is related to the probability $P_{H_1}(Z_o)$ in the following way

$$P_{H_o}(Z_o) = P_{H_1}(-Z_o) = P(Z_o) \tag{60}$$

Using expression (36) for $P(Z_o)$ and applying the constant correspondence given by the table

$P(Z_o)$	$P(-Z_o)$
$K(Z_o)$	$M(Z_o)$
s_1	t_1
m_i	n_i
n	m
ρ_i	φ_i

we obtain the following expression for the probability $P(-Z_o)$

$$\lim_{\lambda\to\infty} P(-Z_o) = M(Z_o)\cdot\lambda^{-t_1}\exp\left(-\frac{\lambda^2}{2}\sum_{i=1}^{m} n_i\varphi_i^2\right)\cdot\exp\left(-\frac{\lambda^2}{2}N\right) \tag{61}$$

The optimum rank test is based on the testing of the likelihood ratio $(P_{H_1}(Z_o))/(P_{H_o}(Z_o))$ against a threshold, which in the case of $-\lambda,+\lambda$ alternatives and λ very large is equal to one. Consideration of expressions (36) and (61) in the derivation of this likelihood ratio results to the following equation

$$LR = \frac{K(Z_o)}{M(Z_o)}\lambda^{t_1-s_1}\exp\left(\frac{\lambda^2}{2}\left[\sum_{i=1}^{n} m_i\rho_i^2 - \sum_{i=1}^{m} n_i\varphi_i^2\right]\right) \tag{62}$$

Because of the high value of λ in (62), the decision breaks into two different schemes, depending on the value of the expression in the brackets at the exponent (62). Namely,

1) if $\sum_{i=1}^{n} m_i\rho_i^2 = \sum_{i=1}^{n} n_i\varphi_i^2$, the test is based on the difference t_1-s_1.

   It is specifically given by

$$t_1 - s_1 \underset{H_1}{\overset{H_1}{\gtrless}} 0 \tag{63}$$

2) if $\sum_{i=1}^{n} m_i\rho_i^2 \neq \sum_{i=1}^{n} n_i\varphi_i^2$, then the test takes form

$$\sum_{i=1}^{n} m_i \rho_i^2 \underset{H_o}{\overset{H_1}{\gtrless}} \sum_{i=1}^{m} n_i \varphi_i^2 \tag{64}$$

The probability of error is now defined as

$$P_e = \sum_{Z_o : H_o \text{ is decided}} P_{H_1}(Z_o) \tag{65}$$

and it is, of course, equal to one minus the power of the test. If the test has taken the form in (63), the probability of error P_e consists of terms $k\lambda^{-\gamma}$. Then, from comparison to the probability of error P_{eo} (appendix B), it is evident that the efficiency of the optimum rank test with respect to the optimal parametric test decreases then toward zero. This degenerate case is therefore not practically interesting.

However, if the optimum rank test has decreased to expression (64), the terms in (65) take the form

$$G(Z_o,\lambda) \exp(-\frac{\lambda^2}{2} \sum_{i=1}^{n} m_i[1 - \rho_i^2]) \tag{66}$$

where $G(Z_o,\lambda)$ is a nonexponential function of λ. For λ very large, the predominant term in (65) will be the one with smaller

$$\sum_{i=1}^{n} m_i(1 - \rho_i^2)$$

exponent. That is, if

$$M_o = \sum_{i=1}^{n_o} m_{oi}(1 - \rho_{oi}^2) = \min_{Z_o : H_o \text{ is decided}} \sum_{i=1}^{n} m_i(1 - \rho_i^2) = M(Z_o) \tag{67}$$

is defined, the P_e probability is then given with very good approximation by

$$P_e = G_o(Z_o,\lambda) \exp(-\frac{M_o}{2} \lambda^2) \tag{68}$$

where $G_o(Z_o,\lambda)$ is again a nonexponential function of λ. The key parameter in (68) is M_o. Since H_o is decided if

$$\frac{P_{H_1}(Z_o)}{P_{H_o}(Z_o)} = \frac{P(Z_o)}{P(-Z_o)} < 1 \quad \text{for large } \lambda$$

the Z_o vector that will accomplish M_o, as given by (67), must be such that

$$M(Z_o) > M(-Z_o) \tag{69}$$

The vectors that satisfy (69) are therefore the only candidates for achieving M_o. Let us now suppose that given an arbitrary constant N_1 there is a particular vector Z_o that satisfies (69) and also gives

$$M(Z_o) = N_1 \tag{70}$$

If such a vector could be found for every assigned value N_1, then reducing N_1, the threshold M_o after which (69) is not satisfied would be obtained. If a vector that satisfies both (69) and (70) exists, then it is at least the vector Z_o that minimizes $M(-Z_o)$ under the assumption that $M(Z_o) = N_1$. To have (69) satisfied, the constant N_1 in (70) must be at least $N/2$, where N is the number of data, while at the same time it cannot exceed N. Also, in (67) the m_i's and ρ_i's are such that

$$\sum_{i=1}^{n} m_i = N, \quad 0 \leq 1 - \rho_i^2 \leq 1$$

$m_i > 0$ and $1 - \rho_i^2$ are not all equal to zero. In a sum

$$\sum_{i=1}^{n} m_i(1 - \rho_i^2)$$

characterized by parameters as explained, the minimum is accomplished if $m_1 = N$, $m_i = 0$ for $i \geq 2$. From appendix A it can be concluded that a rank vector $-Z_o$ that actually realizes M_o in (67) is of a form given by

$$-Z_o = \begin{cases} z_{o1} = z_{o2} = \dots = z_{ok} = 1 \\ z_{o(k+1)} = \dots = z_{on} = -1 \end{cases} \tag{71}$$

Then

$$M(-Z_o) = N - \frac{(2k - N)^2}{N} \tag{72}$$

$$M(Z_o) = k$$

and

$$\min M(-Z_o) = N - \frac{(2N_1 - N)^2}{N} \tag{73}$$

Considering (69) that we want satisfied, we finally obtain $N_1 > \frac{3}{4} N$ which is equivalent to

$$M_o > \frac{3}{4} N \tag{74}$$

In appendix B, the probability of error P_{eo} of the optimal parametric test for the same alternatives H_o, H_1 used in this section and λ very large was found to be given by the following expression

$$P_{eo} = C(\lambda)\exp(-\frac{\lambda^2}{2} N_o) \tag{75}$$

where N_o is the number of data used in this case and $C(\lambda)$ a nonexponential function of λ.

If the two probabilities of error in (68) and (75) are required to be equal (equivalent to equal powers), then the following equation should be satisfied

$$\lim_{\lambda\to\infty} \frac{G_o(Z_o,\lambda)}{C(\lambda)} \exp(-\frac{\lambda^2}{2} [M_o - N_o]) = 1 \tag{76}$$

Asymptotically (for large λ) the requirement expressed by (76) can be interpreted as

$$\lim_{\lambda\to\infty} [M_o - N_o] = 0 \tag{77}$$

which, due to (74) is equivalent to

$$\lim_{\lambda\to\infty} N/N_o = 4/3 \tag{78}$$

In (78), N is the number of the data used by the optimum rank test and N_o the corresponding number for the optimal parametric (Neyman-Pearson) test, that are needed to achieve equal power. The meaning of (78) is that the relative efficiency of the optimum rank test with respect to the optimal nonparametric test converges for high signal-to-noise ratio to a number that is bounded from above by 4/3. As pointed out in [15], this efficiency is not nearly as good as the asymptotic relative efficiency of the same test, which is 1, but considering the small number of samples here (short waiting time and short memory), the two numbers become comparable.

B. One-Sample Wilcoxon Test

The Wilcoxon one-sample test consists of testing the sum

$$\sum_{i=1}^{N} i\cdot z_{oi}$$

against a threshold and deciding accordingly, where N is the number of data and z_{oi} the ith component of the rank vector Z_o. If the values of the z_{oi} components are ± 1, the H_o, H_1 hypotheses are the one-dimensional normal alternatives with means $-\lambda, +\lambda$ correspondingly, and the value of λ is large, the $P(Z_o)$ analysis that was presented in section II and was used in subsection A of section III is valid.

The threshold for λ large is equal to zero and the probability of error P_e is then

$$P_e = \sum_{Z_o:\ \sum_{i=1}^{N} i\cdot z_{oi}<0} P(Z_o) \tag{79}$$

There is again a predominant term in (79) which is given by an expression similar to the one in (66) and there is some Z_o rank vector such that

$$\sum_{i=1}^{N} i \cdot z_{oi} < 0 \tag{80}$$

and

$$M_o = M(Z_o) = \min_{Z_o : \sum_{i=1}^{N} i \cdot z_{oi} < 0} \sum_{i=1}^{n} m_i (1 - \rho_i^2) \tag{81}$$

Then, similarly to subsection A of this section

$$\lim_{\lambda \to \infty} P_e = G(Z_o, \lambda) \exp(- \frac{\lambda^2}{2} M_o) \tag{82}$$

where $G(Z_o, \lambda)$ a nonexponential expression of λ .

Reasoning directly related to the one used in subsection A, leads toward the proper Z_o choice that achieves the M_o in (81). Indeed, with the help of appendix A we conclude that this Z_o has the first ℓ components equal to (-1) and the remaining ones equal to (+1). In addition, the z_{oi} components of Z_o should satisfy (80). From that, the inequality

$$\ell > N/\sqrt{2} \tag{83}$$

is obtained. Also, for a Z_o vector as the one described above, it is easily seen that

$$M_o = \sum_{i=1}^{n} m_i (1 - \rho_i^2) = \ell \tag{84}$$

Due to (83) and (84), we obtain

$$M_o > N/\sqrt{2} \tag{85}$$

Comparison of the probability P_e in (82) to the probability P_{eo} in expression (100) of appendix B, and consideration of (85) leads to the conclusion (in a way similar to the one explained in subsection A) that the relative efficiency of the one-sample Wilcoxon test with respect to the optimum parametric test is bounded from below by $1/\sqrt{2}$, for small number of data and large signal-to-noise ratio. Again this efficiency is considerably inferior to the asymptotic relative efficiency of the same test which is equal to $3/\pi$. However, taking the advantages of small samples under consideration, the two numbers are comparable [15].

Selective values of power and efficiency of the one-sample Wilcoxon test were given by Klotz [9] for normal shift alternatives and small samples where the shift value is finite and the data independent. In the calculation of the probability $P(Z_o)$, the recursive expressions (7) and (8) were used as explained in section II. The Z_o components take 0(+1) values and the H_o , H_1 hypotheses correspond to zero mean and mean μ normal alternatives correspondingly. The false alarm rate α is taken equal to $k/2^N$ where k integer and the power of the test is obtained by adding probabilities $P(Z_N)$ over all the possible sequences that lead toward deciding H_1 . Klotz believed his power values to have accuracy of 4 decimal places and were computed for $\alpha \leq .10$, $N = 1(1)10$, and $\mu = 0, (.25)1.5(.5)3.0$. The efficiency calculated by Klotz was relative to the one-sample t-test. For fixed N, α, μ of the Wilcoxon test, the power of the t-test was computed at the same values of α and μ for two different sample sizes $N_1, N_2 (=N_1+1)$ needed to bracket the Wilcoxon power. Linear interpolation with the power gives $N^* = \lambda N_1 + (1 - \lambda)N_2$ $(0 \leq \lambda \leq 1)$ and with efficiency $e_{w,t}(N,\alpha,\mu) = N^*/N$. A table was developed Klotz. As he points out, it can be observed that this table shows consistence with the fact that the sign test and the Wilcoxon test coincide for the smallest false alarm rate $1/2^N$. The efficiency appears also to decrease with μ as was expected from the analysis in section 6 of the paper by Klotz [9]. This analysis is concerned with the far distance relative efficiency of the Wilcoxon test with respect to the t-one-sample test. This analytical method will not be presented in this work, but the method of calculating P(Z) that is presented in subsection B of section II applies in this case also.

C. The One-Sample Normal Scores Test

The one-sample normal scores test consists of testing the sum

$$\sum_{i=1}^{N} z_{oi} E\{t_{Nr_i}\}$$

against a threshold and deciding accordingly. The meaning of this sum is explained in subsection D of section II, where here we have one sample group instead of two and the t_{Nr_i} is the variable appearing in expression (56) if $N = m + n$. We should remind the reader that the normal scores statistic represents the probability $P(Z)$ when the alternatives are shifted normal with relative distance very small. Of course, the distance d of the two-sample case is substituted by the shift μ (the normal mean under the H_1 hypothesis) in the one-sample problems. Klotz [9] developed some tables on the statistics of the one-sample normal scores tests using the recursive expressions (7) and (8) in subsection A of section II to calculate the rank probability. He observed that the number of different values of the normal scores statistic is in general 2^N and they correspond to the 2^N values of the rank vector. These values are symmetrically placed about the expected value $N/(2\pi)^{\frac{1}{2}}$ on the interval $[0, N(2/\pi)^{\frac{1}{2}}]$. In the development of the power of the normal scores test same methods as in the Wilcoxon case were used. A program for evaluating the weights $E\{t_{Nr_i}\}$ was first written. These values are given up to the 5th decimal in a table in [9]. The number of data N there is taken from 1 to 10 in steps of one.

It was noticed that the power properties of the normal scores and Wilcoxon tests are identical for N and false alarm rate α sufficiently small. It was also noticed that the rejection regions for the same two tests coincide for $\alpha \leq .10$ and $N \leq 5$. For $N = 6, 7$ and $\alpha \leq .10$, it coincides at the natural α levels of the Wilcoxon. Finally, the two tests are equivalent for some other N and α values while for small shifts and comparatively large α values the normal scores test appears to be more powerful.

As a last observation, the Wilcoxon test seems to be improving, when the shifts become large reaching eventually power that is higher than the one achieved by the normal scores test.

D. Two-Sample Tests: General Remarks

The concept of optimum rank tests covered in subsection A of this section, is directly extended to the two-sample problems. The test in these problems still consists of testing the likelihood ratio $P_{H_1}(Z_o)/P_{H_o}(Z_o)$ against a threshold C , where the rank probabilities are given by expressions (38) and (39) in subsection C of section II. The hypotheses H_1, H_o here represent the alternatives of the two samples coming from the same distribution and the same samples coming from different distribution correspondingly. If the alternatives are represented by normal independent data with either equal means or means in distance d from each other, the probabilities $P_{H_1}(Z_o)$ and $P_{H_o}(Z_o)$ are both represented by expression (49) in subsection D of section II, with $d = 0$ and $d \neq 0$ correspondingly. For d large, the efficiency of the two-sample optimum rank test with respect to the optimal parametric is the same as presented in subsection A of this section.

Attempts toward the evaluation of two-sample tests have been made by several investigators. Teichroew [22] observed that the hypotheses H_o and H_1 in such tests can be tested by the use of a critical region. The simplest test consists of selecting, in advance, k particular rank vectors to be in the critical region. If one of these vectors occurs, the hypothesis H_1 that both samples came from the same population, is rejected.

The probability of rejecting the hypothesis when it is true (false alarm rate) is given by

$$\alpha = k \frac{m!n!}{(m + n)!} \tag{86}$$

where m and n are the sizes of the two samples. If randomized tests are used, k is non-integer.

To construct an optimum test, we want a test that, as α increases, the rank vectors into the critical region are placed in the order of decreasing probabilities P(Z). That is so because the power of any alternative is obtained by summing up the probabilities P(Z) of all the Z vectors that belong to the critical region. In case that the hypothesis of having two samples from the same normal population is tested against the alternative that the two samples came from two different normal populations that have the same variance but different means with difference d, the optimum rank test for d small is the $C_1(Z)$ test. This test is explained in subsection D of section II and $C_1(Z)$ is given, in particular, by expression (56). To check if the same test $C_1(Z)$ is sufficiently good for arbitrary values of d, it suffices to rank all the possible vectors Z in the sense of decreasing probability under the assumption of small d and investigate if the same ranking is true for larger d's. According to the previous discussion in the present subsection, a positive answer to that would mean that the test $C_1(Z)$ is as powerful for non-small distances d as for small ones. Teichroew [27] realized such an investigation by computing tables of empirical frequencies of rank vectors that are obtained when a sample of m from N(0,1) and a sample of n from N(0,1) and by 1 if they come from N(-d,1). In those tables the $C_1(Z)$ value is also given for every Z, and several m, n and d values have been considered. The rank vectors have been arranged in order of their $C_1(Z)$ values there.

Wherever the $C_1(Z)$ value is the same for two different Z's , the ordering is by probability. It is observed that the probabilities are increasing essentially monotonically in these tables. That indicates that at least for some significance levels, the construction of uniformly most powerful rank order tests might be possible for the hypothesis above.

Milton [13] also studied the behavior of the $C_1(Z)$ test in comparison to the most powerful for several values of m, n and d . He calculated a table ([13], Table A) that included powers for optimum rank, Wilcoxon and $C_1(Z)$ two-sample tests and observed that there were 29 cases in which the nonrandomized test false alarm rate is less than .10 and the most powerful rank test is more powerful than both Wilcoxon and the $C_1(Z)$ tests. Also, in 15 cases the Wilcoxon test was found to be more powerful than the $C_1(Z)$ test and in 6 of these cases the most powerful rank test was more powerful than the Wilcoxon test. In 14 cases the $C_1(Z)$ test was found to be more powerful than the Wilcoxon test and in 3 of these cases the most powerful rank test was more powerful than the $C_1(Z)$ test. In the case of $m = 7$, $n = 6$, $\alpha = 19/1716$, the Wilcoxon test was found to be more powerful than the locally most powerful rank order test $C_1(Z)$ for $d = .2$.

The two-sample t-statistics mentioned in this section are given by the following expression

$$t = \frac{\bar{x}_m - \bar{y}_n}{\left[\frac{(m-1)s_x^2 + (n-1)s_y^2}{m+n-2} \cdot \frac{n+m}{nm}\right]^{\frac{1}{2}}} \qquad (87)$$

where $\bar{x}_m$, $\bar{y}_n$ the sample means and s_x^2, s_y^2 the sample variances of the two samples observed.

E. Two-Sample Normal Scores Test

The normal scores two-sample test $C_1(Z)$ was first developed in subsection D in section II of this chapter for normal shift alternatives.

From expression (57) it is seen that for distance d sufficiently small, $C_1(Z)$ represents the nondeterministic part of the rank probability $P(Z,d)$. Therefore, the locally (small d) most powerful two-sample test that chooses the Z's that maximize the probability $P(Z)$ for normal shift alternatives is actually represented by the $C_1(Z)$ statistic. Numerous statisticians, among whom Hoeffding [8], Terry [25] and Klotz [10] have studied the properties and the behavior of the two-sample $C_1(Z)$ test which is given by

$$C_1(Z_o) = \sum_{i=1}^{N} z_{oi} E_{N,i} \tag{88}$$

where $N = m + n$ the total number of data, z_{oi} the ith component of the rank vector Z_o that takes values 0 or 1, and $E_{N,i}$ the expected value of the ith smallest order statistic from a sample of N standard normal random variables. The statistic $C_1(Z_o)$ has been shown by Chernoff and Savage [3] to have desirable limiting properties when compared with the t-test. The detailed discussion on these properties is not inside the limits of this chapter. However, it is interesting to point out that the t-statistic was found to require more observations in large samples than does the normal scores statistic to attain the same false alarm rank and power. Klotz [10] extended this observation to smaller samples by first computing $C_1(Z)$ distributions under the null hypothesis and by developing the power tables for both the normal scores and the t-tests under normal shift alternatives.

Specifically, Klotz computed the normal scores power for $N \leq 10$ and compared it with the t . For the calculation of the rank probabilities $P(Z)$, the recursive scheme presented in subsection C in section II of this chapter and expressed by equations (42) and (43) was used. In addition, normal shift alternatives were considered. Detailed description of the method is offered in subsection E, section II of this chapter.

Klotz summarized these power results in table IV of his paper. The vectors Z in it are given only for the upper distribution tail and for S representing the normal scores test, the Z's are ordered by decreasing value of it (S). Power values are given for $m = n = 4,5$ and α near .05 with the corresponding values of the t test at the same false alarm rate for comparison. Accuracy is believed by Klotz to be within one unit in the last place. It is observed that for the small values m, n and α the rejection regions for the Wilcoxon optimal rank test and normal scores test are all the same. For such m, n, α values the t-approximation is therefore found to be satisfactory.

The latest known and most complete work on the power of the $C_1(Z)$ test has been done by Milton [13]. In addition to his comparison tables that were covered in subsection D of section III, he developed power tables of $C_1(Z)$ for normal one-sided and two-sided shift alternatives. From his table A, the smallest number T_o such that $P\{C_1(Z) \geq T_o\} \leq \alpha$ was found for given false alarm rate α.

Then the critical region ω_α for the one-sided case was defined such that it contains the rank vectors Z that satisfy the inequality $C_1(Z) \geq T_o$. For the two-sided case the smallest T_o such that $P\{C_1(Z) \geq T_o\} < \alpha/2$ and $P\{C_1(Z) \leq -T_o\} \leq \alpha/2$ was found. The critical region contains then the Z's such that $C_1(Z) \geq T_o$ or $C_1(Z) \leq -T_o$. The one-sided and two-sided tables developed by Milton are number B-3 and B-4 correspondingly in his book. In those the power of the two-sample t test is given for $\alpha = .24, .10, .05, .025, .01, .005$ and $d = .2(.2)1.0, 1.5, 2.0, 3.0,$ and the values given are to 8 decimal places.

Comparison of tables B-3 and B-4 with table A show differences as large as 1 unit in the sixth decimal place, but generally high accuracy.

F. Wilcoxon Two-Sample Test

If Z_o is the vector which has its ith component equal to zero or one according to if the ith smallest observation among two groups of samples belongs to the first or second group correspondingly, then the two-sample Wilcoxon test consists of testing

$$W(Z_o) = \sum_{i=1}^{m+n} iz_{oi} \tag{89}$$

against a threshold C , where z_{oi} is the ith Z_o component and m, n the number of data in the two sample groups. $W(Z_o)$ in (89) can take mostly $m + n + 1$ distinct values. For one-sided Wilcoxon test, the hypothesis of having no difference between location parameters in the two distributions from which the sample groups come is rejected for large values of $W(Z_o)$ if $d > 0$. In the two-sided test, the same hypothesis is rejected for large or small $W(Z_o)$ values. Milton [13] developed power tables for both one-sided and two-sided Wilcoxon tests. The critical region ω_α for the one-sided test was chosen through values of his table A as in the case of subsection E, section III. From table A an integer T_o was found such that $P\{W(Z_o) \geq T_o\} \leq \alpha$ and $P\{W(Z_o) \geq T_o - 1\} > \alpha$, for given α . Then, ω_α includes all Z_o's such that $W(Z_o) \geq T_o$. For the two-sided test the corresponding integer T_o was such that $P\{W(Z_o) \geq T_o\} < \alpha/2$ and $P\{W(Z_o) \geq T_o - 1\} > \alpha/2$, and the critical region includes all Z_o such that $W(Z_o) \geq T_o$ or $W(Z_o) \leq n(n + 1)/2 + mn - T_o$. These tables are numbered B-1 and B-2 in Milton's book and include results for one-sided and two-sided Wilcoxon two-sample tests correspondingly. In them below each power value of the Wilcoxon test on the line labeled t , is given the power of the two-sample t-test for the same α and shift alternatives. Also, the α and d values used in these tables are the same with the ones used in subsection E, section III. The form of the test $W(Z_o)$ in the two-sample case does not differ from its form in the one-sample case. The methods for power computation are, therefore, the same in both cases. Furthermore, the analytical results presented in subsection B, section III for far distance normal alternatives apply here directly. Finally, tables exhibited by several authors on the one-sample Wilcoxon test (as covered in Section 3b) can compare with the tables (B-1) and (B-2) in [13] that were obtained for the two-sample test.

G. Conclusions

The small sample power of some of the most characteristic nonparametric tests was studied in this section. The analytical methods and the computational results presented come exclusively from independent, normal data and shift alternatives. The effect of dependence on the behavior of the several rank tests has been studied by some investigators [5,7,15,27, 28] only under the assumption of large samples. It has been found that then these tests stop being distribution free and their efficiencies and powers are considerably different than what they were when the data were independent. It is expected that the small-sample behavior of such rank tests will also be considerably different when independence is not a good assumption.

The optimal rank, the Wilcoxon and the normal scores one-sample tests were covered in this section. The efficiency of the first two with respect to the optimal parametric test was found analytically for far distance normal alternatives. The results in [9] include efficiencies of the Wilcoxon test with respect to the t and for normal shift alternatives. The efficiency is observed to decrease with the shift, which agrees with the analytical results for large shift. From the power tables in [9] it is observed that the normal scores and Wilcoxon one-sample tests take terms in achieving power superiority for false alarm rate and sample number varying.

Similarly, as summarized in table III in [9], the most powerful, Wilcoxon, normal scores and t two-sample tests, each also achieves power superiority for different false alarm and sample number values.

IV. SEQUENTIAL TWO-SAMPLE TESTS AND SELECTION PROCEDURES

The problems involving small-sample rank tests are of such variety that they cannot be covered all extensively in this chapter. It is the opinion of the author, however, that the reader of this book would get an incomplete idea of the use of the rank procedures if the sequential two-sample tests and some selection procedures were not mentioned shortly.

The purpose of the sequential two-sample tests is to discriminate between two hypotheses using the ranked data from two sample groups. The null hypothesis $H_o(H_1)$ corresponds to having both sample groups from the same (different) distributions. The problems that have been solved involve independent normal data sequentially received and shift alternatives. At each stage, (m + n) observations are taken consisting of m from one distribution and n from the other. Then, the two-sample rank vector with 0(1) components is formed, as explained in section II and the probabilities $P_{H_1}(Z)$ and $P_{H_o}(Z)$ (conditioned on the two hypotheses correspondingly) are expressed. The probability ratio r_t when t groups have been observed, is given by

$$r_t = \prod_{i=1}^{t} p_i \tag{90}$$

where

$$p_i = \frac{P_{H_1}(Z_i)}{P_{H_o}(Z_i)} \tag{91}$$

and Z_i the rank vector of the ith double group. If α is the desired false alarm rate and β the desired power, the ratio r_t is tested against the two constants $A = (1 - \beta)/\alpha$ and $B = \beta/(1 - \alpha)$ and a decision is taken accordingly. Namely,

1) if $B < r_t < A$, one more double sample is taken
2) if $r_t \geq A$, a decision is taken in favor of hypothesis H_1
3) if $r_t \leq B$, a decision is taken in favor of hypothesis H_o

The number of possible Z_i vectors for every group i is given by $\binom{m+n}{n}$. In addition, under the hypothesis H_o , all the rank vectors are obviously equally likely. Therefore, $P_{H_o}(Z_i) = 1/\binom{m+n}{n}$ and from (90) we obtain

$$p_i = \binom{m+n}{n} P_{H_1}(Z_i) \tag{92}$$

If shift alternatives with distance d are involved, the probability ratio (with substitution of (92)) is given by the expression

$$r_t = \binom{m+n}{n}^t \prod_{i=1}^{t} P(Z_i/d) \tag{93}$$

The two-sample test expressed by (93) is known as the CSR test (sequential, configural rank test). A different version of the two-sample sequential test with shift alternatives is expressed by the SRS test (sequential, rank sum test) and is given by

$$r_t = \prod_{i=1}^{t} \frac{P(W_i/d)}{P(W_i/o)} \tag{94}$$

where

$$w_i = \sum_{j=1}^{m+n} jz_{ij} \tag{95}$$

and z_{ij} is the ith component of the rank vector Z_{i1} that represents the one of the two samples (with number of data n) taken at the instant i.

Both the SCR and the SRS tests terminate with probability one. Also, the operating characteristic function and the average sample number function are the two concepts that have been used for evaluation of these tests. The first gives the probability of accepting H_o when d is the true value of the shift in location between the two populations, and the second gives the average sample sizes required for termination of the test for various true values of d. Milton [13, Ch. 3] provides tables that give values of these concepts for the SCR and SRS tests.

The goal of the selection procedures is the selection of a subset that contains at least a certain minimum number of the "best" among k populations. The "best" is defined by the value of a certain characteristic, which for the problems that have been studied is equal to the statistical mean of each population. Specifically, it is assumed that k populations of independent, usually normal, random variables X_i are observed such that the variance is common, known and equal to σ^2, and the mean μ_i of each population is unknown. Milton [13, Ch. 4] denotes

$$\mu_{[1]} \leq \mu_{[2]} \leq \cdots \leq \mu_{[k]}$$

the ranked means and

$$\bar{X}_{[1]} \leq \bar{X}_{[2]} \leq \cdots \leq \bar{X}_{[k]}$$

the ranked sample means of the populations. The association of each of the populations with a certain mean μ is not known to the observer. The goal is to select a subset of populations of size s that contains at least c of the t best populations, with probability of correct selection $p_o (p_o < 1)$. In addition, to make the problem solvable, we want to be able to find such a scheme whenever the distance between $\mu_{[k-1]}$ and $\mu_{[k-t+1]}$ is at least equal to a nonzero given constant d_o. Mahamunulu [12] has shown that to achieve this goal, N observations from each population must be taken, then the sample means $\overline{X}_1, \cdots, \overline{X}_k$ must be computed and ranked and the populations associated with the s largest sample means must be selected.

In addition, Mahamunulu also found a lower bound for N, if fixed p_o and d_o constants are given. This bound varies according to whether the subset of size s has to include the t best populations $(s \geq t)$ or any s of the t best populations $(s \leq t)$.

Milton [13] derived tables that can be used for solving the selection procedures presented by Mahamunulu. These tables are based on the tables including the $P(Z)$ values and they are explained in chapter 4 of Milton's book.

Conclusion

The sequential tests and the selection procedures are not in the limits of the material that is covered in this book and they are covered only shortly as applications of the rank probability $P(Z)$. For the interested reader, we recommend the book by Milton [13] for further study. In chapters 3 and 4, he covers these two subjects and he offers tables and complete bibliography.

V. SUGGESTIONS FOR FURTHER RESEARCH

The small-sample studies of the nonparametric tests have been restricted in cases of normal independent data and shift alternatives. The reason for that is that there are mathematical complications involved otherwise. However, the dependence of the samples is too important to be left unexploited. Indeed, in many cases the sampling distance is not large enough for the samples to be considered independent. As was mentioned in the conclusion of section III, the influence of dependence on the behavior of the nonparametric tests has been studied only in the asymptotic case of large samples [5,7,15,27,28]. It is claimed here that some of the asymptotic ideas can be applied to small samples and that some kinds of dependence (such as the first order Markov) can be handled numerically in an efficient way.

Another interesting extension of the small-sample studies consists of considering more generalized alternatives. For example, the checking of the variances of the received samples might be desirable instead of the checking of the means. The behavioral study of the known nonparametric tests in this case is very interesting for normal independent as well as normal dependent data.

If penalties due to time delays are introduced, then in addition to the power, the delay costs should also be considered for the evaluation of rank tests under given sample size. The problem becomes more complex in this case and tables including the test powers as well as the delay penalties need to be formed. The choice of the proper sample size for given rank test or the choice of the proper test for given sample size will depend then on a certain function of both power and delay penalty.

An interesting problem related to the sequential decision schemes is the one of considering delay penalties in forming stopping rules for the rank tests. The problem is related to section IV of this chapter and the stopping times are varying with the different choices of delay costs. The introduction of such costs to the rank problems is very important adding one more realistic factor to the whole problem. In many real cases, the time that can be afforded by the system for accumulating the data and making a decision is limited and the delay risks involved very high.

ACKNOWLEDGMENT

This work was supported by the National Science Foundation under Grant ENG 75-10516 and Air Force Grant AFOSR 77-3156.

APPENDIX A. DYNAMIC PROGRAMMING METHOD FOR THE CALCULATION OF X_{max}

The objective here is the calculation of the vector X_{max} that maximizes the product

$$\prod_{i=1}^{N} \Phi(x_i - z_{oi}\lambda) \tag{96}$$

where the variables t_i are positive and ordered so that t_i is also the ith smallest. Also, z_{oi} are given constants equal to either +1 or -1, and they are taken to be the components of an N-dimensional vector Z_o. The vector X_{max} will be found through dynamic programming methods and its value will, of course, depend on the particular vector Z_o used.

The problem breaks down to finding the minimum with respect to the x_i of the expression

$$\sum_{i=1}^{N} (x_i - z_{oi}\lambda)^2$$

under the restriction of all x_i being positive and ordered. We proceed by stating the following lemma.

Lemma 1: If Z_o is such that $z_{o1} = \dots = z_{ok} = -1,\ k \geq 1$, then it is such that its first k components $x_{1max},\dots,x_{kmax}$ are all equal to zero.

Proof: Using dynamic programming, we fix the values of $x_2,\dots,x_N$ and minimize the expression

$$\sum_{i=1}^{N} (x_i - z_{oi}\lambda)^2 = \sum_{i=1}^{N} (x_i + \lambda)^2 + \sum_{i=k+1}^{N} (x_i - z_{oi}\lambda)^2$$

with respect to x_1. That gives $x_1 = x_{1max} = 0$. Proceeding similarly, we get the result in the statement of the lemma.

Lemma 2: Let the k first components of the vector Z_o be equal to +1, followed by μ components equal to -1. Then, X_{max} is such that

$$x_{1max} = x_{2max} = \dots = x_{k+\mu,max} = 0 \ ,$$

$$\text{if} \quad \mu \leq k$$

$$x_{1max} = x_{2max} = \dots = x_{k+\mu,max} = \frac{k - \mu}{k + \mu} \lambda$$

$$\text{if} \quad \mu < k \text{ and } x_{k+\mu+1,max} \leq \frac{k - \mu}{k + \mu} \lambda$$

$$x_{1max} = x_{2max} = \dots = x_{k+\mu,max} = x_{k+\mu+1,max}$$

$$\text{if} \quad \mu < k \text{ and } x_{k+\mu+1,max} < \frac{k - \mu}{k + \mu} \lambda$$

Proof: In the case of a Z_o vector as described in the statement of the lemma, the sum

$$\sum_{i=1}^{N} (x_i - z_{oi}\lambda)^2$$

that we want to minimize with respect to the x_i is equal to

$$\sum_{i=1}^{k} (x_i - \lambda)^2 + \sum_{i=k+1}^{k+\mu} (x_i + \lambda)^2 + \sum_{i=k+\mu+1}^{N} (x_i - z_{oi}\lambda)^2 \tag{97}$$

Using dynamic programming methods, we see that if $x_2, x_3, \dots, x_N$ are kept constant, the x_{1max} that minimizes the expression in (97) under the restriction of $x_1 \leq x_2$ is then equal to x_2. We note here that to every x_i such that $\lambda < x_i \leq 2\lambda$ corresponds an x_i' such that $0 \leq x_i' \leq \lambda$ and $(x_i - \lambda)^2 = (x_i' - \lambda)^2$. Hence we will assume here that the x_i that minimize the first partial sum in (97) are all in the region $[0,\lambda]$.

Let us now suppose that the values x_{1max}, x_{2max} up to x_{kmax} are determined. By the dynamic programming logic, they will all be equal to each other. The question is, what is the value $x_{k+1,max}$ that achieves the minimum of (97) with respect to x_{k+1} under the restriction $x_{k+1,max} \geq x_{k,max}$? The answer is obviously $x_{k+1,max} = x_{k,max}$.

Finally, given the value $x_{k+1,max}$ and, since the terms $(x_i + \lambda)^2$ are minimum for the smallest possible positive x_i values, then under the restriction $x_{k+1,max} < x_{k+2,max} \leq \ldots \leq x_{k+\mu,max}$, the minimum of the second partial sum in (97) is assumed for $x_{k+1,max} = x_{k+2,max} = \ldots = x_{k+\mu+1,max}$. From what has already been said in this proof, it is clear that the sum in (97) becomes equal to

$$k(t_o - \lambda)^2 + \mu(x_o + \lambda)^2 + \sum_{i=k+\mu+1}^{N} (x_i - z_{oi}\lambda)^2 \tag{98}$$

where x_o is the common value that all variables $x_1, \ldots, x_{k+\mu}$ assume in the minimum. To actually find this value x_o, we differentiate (98) with respect to it and put the result equal to zero. The second derivative is always $k + \mu > 0$ so the extreme, if it is in the region $[0,\lambda]$, is the minimum.

Indeed, the first derivative gives $x_o = \lambda(k - \mu)/(k + \mu)$. If $k - \mu \geq 0$, t_o is positive and it assumes the minimum under the restriction $x_1 \leq \ldots \leq x_N$. If $k - \mu < 0$, however, the value $\lambda(k - \mu)/(k + \mu)$ is outside the acceptable region. But for $x_o > (k - \mu)/(k + \mu)$, (98) increases monotonically, assuming the minimum, under the restriction $x_o > 0$, at $x_o = 0$.

By extending the results in this lemma in a more-or-less straightforward manner, we reach the following lemma.

Lemma 3: Let the components of vector Z_o be grouped in the following way:

The ith group consists of k_i consecutive Z_o components of value equal to +1, followed by μ_i consecutive components of value equal to (-1). Then, the following results are true.

1) if $k_i > \mu_i$, and at the same time $k_i/\mu_i \leq k_{i+1}/\mu_{i+1}$, for all

i , then X_{max} is such that

$$x_{j,max} = \frac{k_1 - \mu_1}{k_1 + \mu_1} \lambda , \quad j = 1,2,\ldots,k_1 + \mu_1$$

$$x_{j,max} = \frac{k_i - \mu_i}{k_i + \mu_i} \lambda , \quad j = \sum_{i=1}^{i-1} (k_i + \mu_i),\ldots, \sum_{i=1}^{i} (k_i + \mu_i)$$

2) if $k_i > \mu_i$, for all i , and, at the same time, $k_j/\mu_j \le k_{j+1}/\mu_{j+1}$, for $j = 1,2,\ldots,i-1$, while

$$\frac{k_i}{\mu_i} > \frac{k_{i+1}}{\mu_{i+1}}$$

then we find the smallest integer L such that

$$\frac{\sum_{m=0}^{L+1} k_{i+1-m}}{\sum_{m=0}^{L+1} \mu_{i+1-m}} \le \frac{k_{i-L-1}}{\mu_{i-L-1}}$$

and we have the same results as in case 1) under the new groups $(k_1 + \mu_1)$,

$$(k_2 + u_2),\ldots,(k_{i-L-1} + \mu_{i-L-1})$$

$$(\sum_{m=0}^{L-1} k_{i+1-m} + \sum_{m=0}^{L-1} \mu_{i+1-m})$$

3) if $k_i < \mu_i$, for some i , then again we find the smallest integer L such that

$$\frac{k_{i-L-1}}{\mu_{i-L-1}} \ge \frac{\sum_{m=0}^{L+1} k_{i-m}}{\sum_{m=0}^{L} \mu_{i-m}}$$

and proceed as in case 2.

4) if

$$(\sum_{j=1}^{i} k_j / \sum_{j=1}^{k} \mu_j) \le 1$$

for some i , then $t_{j,max} = 0$, for

$$j = 1,2,\ldots, \sum_{i=1}^{i} (k_i + \mu_i)$$

5) if the last group of Z_o components has all its members equal to +1 , then X_{max} components corresponding to this group are all equal to λ .

Summarizing the results in this appendix, we have

$$\max_{x_{i,i+1,\ldots,N}} \prod_{i=1}^{N} \Phi(x_i - z_{oi}\lambda) = \prod_{i=1}^{n} \Phi^{k_i}(\xi_i\lambda) \cdot \Phi^{\mu_i}(\rho_i\lambda)$$

under the restriction $0 \leq x_{1,max} \leq x_{N,max}$, where

n number of groups of Z_o components

m_i number of components in the ith group

k_i number of +1 components in the ith group

μ_i number of -1 components in the ith group

$$\xi_i = \begin{cases} 1, & \text{in case 4} \\ \dfrac{m_i - \Sigma_j z_{oj}}{m_i}, & \text{in cases 1 to 3} \end{cases}$$

$$\rho_i = \begin{cases} 1, & \text{in case 4 and Lemma 1} \\ \dfrac{m_i + \Sigma_j z_{oj}}{m_i}, & \text{in cases 1 to 3} \end{cases}$$

$\Sigma_j z_{oj}$ summation of all Z_o components in the ith group

APPENDIX B. PROBABILITY OF ERROR OF THE OPTIMUM PARAMETRIC TEST

Two hypotheses are given as follows:

H_1 : observations Gaussian, with mean λ and unit variance

H_o : observations Gaussian, with mean $-\lambda$ and unit variance

Take N_o observations $\{x_i\}$ independent from each other to test the two hypotheses. The optimal parametric test is (as well known) given by

$$T(X) = \frac{1}{N_o} \sum_{i=1}^{N_o} x_i \underset{H_o}{\overset{H_1}{\gtrless}} C \tag{99}$$

where C is the threshold and T(X) is Gaussian with variance $1/N_o$ and means λ or $-\lambda$ depending on the hypothesis under which it is examined.

For λ very large, and $C = 0$, the probability of error P_{eo} is given by

$$P_{eo} = P\{\text{decide } H_o/H_1\} = \int_{-\infty}^{o} \frac{1}{\sqrt{2\pi}(1/\sqrt{N_o})} \exp(-\tfrac{1}{2}N_o[u - \lambda]^2)du =$$

$$= \int_{-\infty}^{o} \frac{1}{\sqrt{2\pi}(1/\sqrt{N_o})} \exp(-\tfrac{1}{2}N_o[u + \lambda]^2)du =$$

$$= \int_{-\infty}^{-\lambda\sqrt{N_o}} \varphi(u)du \underset{\lambda\to\infty}{\to} \frac{\varphi(-\lambda\sqrt{N_o})}{\lambda\sqrt{N_o}} = \frac{\exp(-\frac{\lambda^2}{2} N_o)}{\sqrt{2\pi N_o}\,\lambda} =$$

$$= C(\lambda) \exp(-\frac{\lambda^2}{2} N_o) \tag{100}$$

REFERENCES

1. Ahrens, E.H., Sr. (1969) "Mass field trials of the diet-heart question (their significance, timeliness, feasibility and applicability)," American Heart Association Monograph No. 28.

2. Berk, R.H. and I. Savage (1968) "The information in a rank-order and the stopping time of some associated SPRT's," Ann. Math. Stat. 39: 1661-1674.

3. Chernoff, H. and I.R. Savage (1958) "Asymptotic normality and efficiency of certain nonparametric test statistics," Ann. Math. Stat. 29: 972-994.

4. Davisson, L.D., E.A. Feustel, and J.W. Modestino (1970) "The effects of dependence on nonparametric detection," IEEE Trans. Inf. Th. IT-16: 32-41.

5. Davisson, L.D. and P. Papantoni-Kazakos (1972) "Small sample efficiencies of rank tests," Proc. 1972 International Symp. on Inf. Th., p.57.

6. Gaswirth, J.L. and H. Rubin (1971) "Effect of dependence on the level of some one-sample tests," J. Amer. Stat. Assoc. 66: 816-820.

7. Hodges, J.L. and E.L. Lehmann (1962) "Probabilities of rankings for two widely separated normal distributions," in Studies in Mathematical Analysis and Related Topics (Gilbarg, Solomon, et al., eds.), Stanford University Press, Stanford, California, pp. 146-151.

8. Hoeffding, W. (1951) "Optimum nonparametric tests," Proc. Second Berkeley Symp. Math. Stat. and Prob. (J. Neymann, ed.), University of California, Berkeley, pp. 83-93.

9. Klotz, J.H. (1963) "Small sample power and efficiency for the one-sample Wilcoxon and normal scores test," Ann. Math. Stat. 34: 335-337.

10. Klotz, J.H. (1964) "On the normal scores two-sample tests," J. Amer. Stat. Assoc. 59: 652-664.

11. Lehmann, E.L. (1953) "The power of rank tests," Ann. Math. Stat. 24: 23-42.

12. Mahamunulu, D.M. (1967) "Some fixed-sample ranking and selection problems," Ann. Math. Stat. 38: 1079-1091.

13. Milton, R.C. (1970) Rank Order Probabilities: Two-Sample Normal Shift Alternatives. J. Wiley & Sons, Inc., New York.

14. Modestino, J.W. (1969) Nonparametric and Adaptive Detection of Dependent Data. Ph.D. Dissertation, Dept. of Elec. Engr., Princeton Univ., Princeton, N.J.

15. Papantoni-Kazakos, P. (1975) "Small sample efficiencies of rank tests," IEEE Trans. Inf. Th. IT-21: 150-157.

16. Puri, M.L. (ed.)(1970) Nonparametric Techniques in Statistical Inference. Cambridge Univ. Press, London, England.

17. Rao, U.V.R., I.R. Savage, and M. Sobel (1960) "Contributions to the theory of rank-order statistics: The two-sample censored case," Ann. Math. Stat. 31: 415-426.

18. Savage, I.R. (1956) "Contributions to the theory of rank order statistics -- the two-sample case," Ann. Math. Stat. 27: 590-615.

19. Savage, I.R. (1957) "Contributions to the theory of rank order statistics -- the 'trend' case," Ann. Math. Stat. 28: 968-977.

20. Savage, I.R. (1960) "Contributions to the theory of rank-order statistics: Computation rules for probabilities of rank orders," Ann. Math. Stat. 31: 519-520.

21. Smirnov, N.V. (1948) "Table for estimating the goodness of fit of empirical distributions," Ann. Math. Stat. 19: 279-281.

22. Teichroew, D. (1955) "Empirical power functions for nonparametric two-sample tests for small samples," Ann. Math. Stat. 26: 340-344.

23. Teichroew, D. (1955) Probabilities Associated with Order Statistics in Samples from Two Normal Populations with Equal Variance. Engineering Agency Statistical Tables. Chemical Corps Engineering Agency, Army Chemical Center, Maryland.

24. Teichroew, D. (1956) "Tables of expected values of order statistics and products of order statistics for samples of size twenty and less from the normal distribution," Ann. Math. Stat. 27: 410-426.

25. Terry, M.E. (1952) "Some rank order tests which are most powerful against specific parametric alternatives," Ann. Math. Stat. 23: 346-366.

26. Tsao, C.K. (1957) "Approximations to the power of rank tests," Ann. Math. Stat. 28: 159-172.

27. Woinsky, M.N. and L. Kurz (1970) "Nonparametric detection using dependent samples," IEEE Trans. Inf. Th. IT-16: 355-358.

28. Wolff, S.S., J.L. Gastwirth, and H. Rubin (1967) "The effect of autoregressive dependence on nonparametric test," IEEE Trans. Inf. Th. IT-13: 311-313.

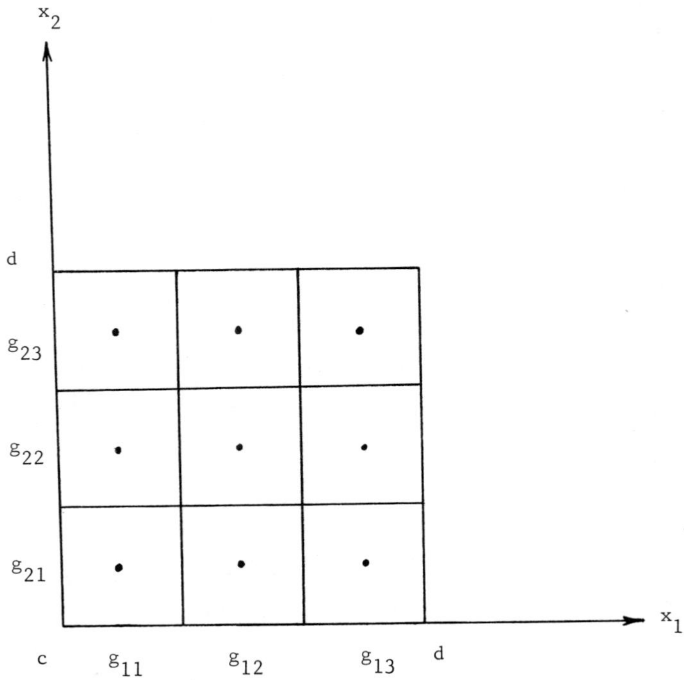

Figure 1 Region of integration for (10)

Z_{oN}

Z_{o1} : (1)		(0)	
Z_{o2} : (1,1)		(0,1)	(0,0)
Z_{o3} : (1,1,1) → (1,1,1)	(1,0,1) → (1,0,0)	(0,1,1) → (0,1,0)	(0,0,1) → (0,0,0)

Figure 2 Order of Computations $N_{max} = 3$

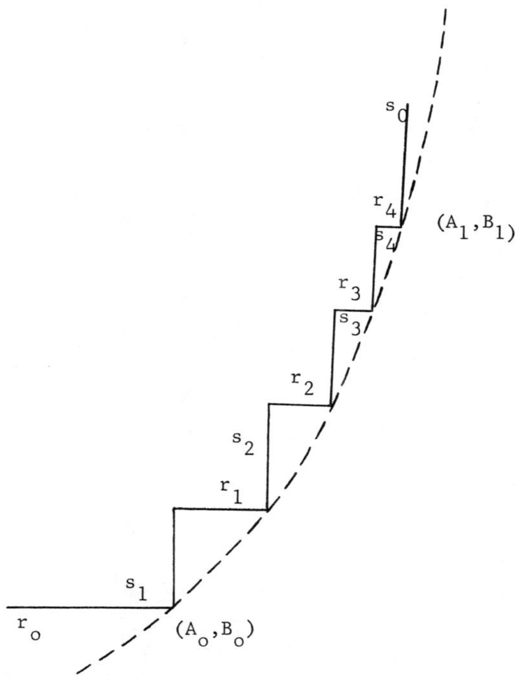

Figure 3

Chapter 6

ON NONPARAMETRIC ESTIMATION OF PROBABILITY DENSITY FUNCTIONS

Dimitri Kazakos

Department of Electrical Engineering
State University of New York at Buffalo
Buffalo, New York

I. INTRODUCTION

In several aspects of statistical data processing, one of the most fundamental assumptions frequently used is the knowledge of the probability law that governs the generation of the data. Within the framework of probability laws, the most fundamental component is the probability density function (PDF).

Statistical modeling of the data has been a successful approach in the areas of communication, image processing, pattern recognition, decision theory, estimation, stochastic control and others. In all of the above areas, optimal processing of the observed data requires knowledge of the probability density function for each statistical population. For example, in detecting a signal in noise, the Neymann-Pearson (likelihood ratio) test is the optimal one. To implement it, we must know the probability density functions under both hypotheses. The very reason for considering nonparametric statistical tests is our lack of knowledge of the probability density functions. In other words, we accept de facto our lack of knowledge of the environment and proceed to design a system that will perform acceptably in spite of our incomplete knowledge.

In the present chapter we adopt a different philosophy. We assume that the situations are such that we can afford to "learn" the statistics of the environment. Clearly, if the resources for such a task are available, the resulting system will perform better than the nonparametric one, because complete statistical knowledge of the environment will be available to the designer.

An intermediate situation is presented in Chapter 3 by Kurz. His partition detectors require knowledge only of the quantiles of the corresponding probability density functions. His approach requires much less complexity than a full-fledged density estimator.

In data communication situations where the transmitter and receiver complexities are limiting factors, a simple estimator is preferable. However, integrated circuit technology and other technical advances tend to push the "complexity ceiling" higher. As Lucky [13] commented, in view of increased processing capabilities, we should think of transmitters and receivers as computer programs and algorithms rather than electronic devices.

In other situations where larger computer memory and processing complexity is affordable due to the off-line character of the computations, as is the case in statistical pattern recognition, use of more sophisticated density estimators is unquestionably warranted. As an example we cite the automatic processing of biomedical data. A number of measurements of the vital signs of a patient constitutes a k-dimensional observation vector x . We model the healthy state of a patient as hypothesis H_1 and his sick state as H_2 . If the probability density functions of x under H_1 and H_2 are accurately known, the computer can decide optimally on the patient's state by a likelihood ratio test. In the above example it is clearly worthwhile to estimate the densities as well as possible at the expense of greater complexity.

If we have sufficient past experience on data of similar origin, we can frequently constrain the form of the underlying density as being a member of a family of functions $f(x/q)$ indexed by an r-dimensional parameter q . In such a case, estimation of the density is equivalent to the estimation of the true value of q on the basis of a number of observations $(x_1,\ldots,x_n)$. We are then led to parameter estimation theory and methods. The structure of f is rigid and invariant for the new observations. Parameterization of the density implies previous "peeking" at similar data for assessing the structure to be used.

An alternative to the rigidity and "previous experience" required for using parametric families of density estimators is the class of nonparametric ones. In contrast to nonparametric detection, there is no strict definition of a nonparametric density estimator. A qualitative requirement must be satisfied by an acceptable nonparametric density estimator, namely that its structure and performance depend minimally on any previous knowledge of the nature of the true underlying probability density function. A usual and mild requirement is the smoothness of the true PDF measured by the number of its existing derivatives. Since there is no strict definition of nonparametricness of a PDF estimate, the thinking behind the construction of one has to be heuristic by necessity. Of course the analysis of its performance must be mathematically rigorous.

Density estimation has been a subject of great interest to pure and applied statisticians as well as to systems scientists involved in pattern recognition. Members of the second group frequently miss statistical results of potentially great practical applicability. It is the purpose of the present chapter to bring to attention several nonparametric density estimation results together with the author's derivation of the optimal kernel function for kernel type density estimates. Also several open questions of practical and theoretical importance will be outlined. In a pair of articles, Cover [3] and Wegman [29] surveyed very thoroughly the state of the art in density estimation. In the present chapter an updating review of some density estimation methods will be attempted together with the author's new result. The latter consists in finding the optimal kernel function for Parzen's kernel type density estimates. Its form depends minimally on knowledge about the true density. Only the degree of smoothness of the density is required known and this is clearly a nonparametric piece of information. The practical advantages and disadvantages of each density estimator will be discussed in terms of computational requirements, small sample and asymptotic performance. Open problems of theoretical and practical significance will be also proposed.

The list of references given at the end covers only a very small subset of the work in the area. We only mention those references that are closely related to the results reported in the present chapter.

II. PERFORMANCE MEASURES

The performance of density estimators has been assessed by means of several convergence criteria. The extent of assumptions required for establishing each kind of convergence is varying and so is the strength of convergence results obtained. The amount of assumptions and type of convergence established provide us with a varying degree of faith in each estimator for a particular situation.

Let $f_n(x)$ be an estimate of the probability density function $f(x)$ based on n samples $(x_1,\ldots,x_n)$. Then the expected value $E[f_n(x) - f(x)]^2 = M_n(x)$ averaged over the n samples and for a fixed x is called the mean square error. If $M_n(x) \to 0$ as $n \to \infty$, we say we have pointwise consistency in quadratic mean. If $M_n(x) \to 0$ as $n \to \infty$ uniformly in x, then we call it uniform consistency in quadratic mean. If $E\int [f_n(x) - f(x)]^2 dx \to 0$ as $n \to \infty$, we say we have integratedly uniform consistency in quadratic mean. If

$$E[\sup_x |f_n(x) - f(x)|^2] \to 0$$

for $n \to \infty$, we say that the estimate is functionally uniformly consistent in quadratic mean. Aside from the mean square concept, other types of consistency may be defined by convergence in probability and with probability one. Uniform consistency in probability implies that

$$\Pr[\sup_x |f_n(x) - f(x)| < \epsilon]$$

converges to 1 for all $\epsilon > 0$. By pointwise consistent with probability one we mean $|f_n(x) - f(x)| \to 0$ with probability one, and by uniformly consistent with probability one that

$$\sup_x |f_n(x) - f(x)| \to 0$$

with probability one. Clearly, uniform consistency is preferable to pointwise and consistency with probability one is preferable to consistency in quadratic mean or in probability. Also consistency "with probability one" is better than "in probability" or "in quadratic mean."

The preceding comparative list of types of convergence follows Wegman [29] and is included for completeness.

In establishing the convergence properties of a density estimator, we typically attempt to ensure the strongest type of convergence possible using the fewest assumptions on the probability density function that will do it. The preference of some types of convergence may be frequently dictated by the application for which we intend to use the estimator. For example, if the density estimate is to be used for subsequent classification of new data, the convergence criteria based on

$$\sup_x |f_n(x) - f(x)|$$

are much more relevant for the task because they guarantee a pointwise fit that results in a likelihood ratio decision rule performing well in all of the observation space.

III. MAXIMUM LIKELIHOOD (ML) ESTIMATES OF DENSITY FUNCTIONS

The most straightforward and immediate density estimate that comes from the definition of the probability density function is the histogram. It is widely used by scientists. However, its many shortcomings have prompted the consideration of alternative estimators that perform better both theoretically and practically. The main objection to the histogram is its lack of smoothness. In most real problems, physical considerations imply that the true underlying probability density function is smooth, while the histogram is not. Consequently, some type of smoothing of the data is always desirable.

In designing a density estimator there is clearly some intuition and arbitrariness involved. We can impose the restriction that the true PDF belongs to a parametric family of functions and our task then is reduced to one of parameter estimation. In this case, the desired "smoothing" of the data is implicitly induced by the parametric structure imposed. Clearly the analysis of performance of the resulting parameter estimation problem is also simpler.

A great shortcoming limits the usefulness of parametric density estimation, however. By parametrizing the class of densities, we make it very small. Essentially we imply that we already know a lot about the true density. Parametric structures generally lack in flexibility of adaptation.

The term "nonparametric" ideally implies an estimator or procedure in general that is valid or performs well for any true underlying probability density function. Invariably there is a need to restrict in some sense the class of densities for assessing the performance of the nonparametric estimator. As Wegman [29] succinctly put it, the criterion for nonparametricness is the cardinality or size of the family of densities for which the procedure is successful.

Smoothing and filtering of the data is an important concept that appears in many statistical situations. In estimating the spectrum of a random process we have the situation of "rough" data while we have reasons to believe in the smoothness of the underlying spectrum. Hence several smoothing filters or windows have been employed [11]. In linear least squares estimation of a random process in noise, the concept of data smoothing appears again. When a finite number of samples is given and we want to estimate the underlying probability density function, we could presumably consider it to be a sum of Dirac delta functions located at the data samples and with total mass equal to one. However, this is a very bad estimate of a smooth probability density. Smoothing of the data before defining any estimator is very desirable and intuitively satisfying. Such an idea of data smoothing or smearing prevails in the construction of kernel estimates.

Maximum likelihood parameter estimation methods have been widely studied and used. The first to propose the maximum likelihood procedure for parametric density estimation was R.A. Fisher [9]. We will summarize here recent extensions of the maximum likelihood density estimation procedure in function spaces of higher cardinality. In other words, the search is not in a finite dimensional parameter space, but in a function space with consequent achievement of greater flexibility and adaptability to the available data.

Let S denote the interval $[a,b]$. We consider the problem of estimating the probability density function $f \in L^1(S)$ that generates a sequence of independent random variables

$$x_1,\ldots,x_n \in S$$

Define $L(v)$, the likelihood that a probability density function $v \in L^1(S)$ generated the random sample $x_1,\ldots,x_n$ by:

$$L(v) = \prod_{i=1}^{n} v(x_i)$$

Let $H(S)$ be a manifold is $L^1(S)$ and consider the following optimization problem:

$$\begin{aligned}&\text{maximize} \quad L(v) \; ; \\ &\text{under the constraints} \\ &v \in H(S) \; , \int_S v(t)\,dt = 1 \\ &v(t) \geq 0 \;\forall\; t \in S\end{aligned} \tag{1}$$

By a maximum likelihood estimate based on the sample $x_1,\ldots,x_n$ we mean any solution of the constrainted optimization problem (1). If $H(S)$ is a finite dimensional manifold, we have a parametric estimate. Otherwise, we have a nonparametric one. One of the difficulties we may encounter in the nonparametric case is that the maximum value of L may be infinite and achieved for a $v(t)$ that is a sum of Dirac delta functions if the structure of $H(S)$ permits this to happen. Such a solution is unacceptable and we should structure the problem in a way that avoids it.

One possibility is to choose the infinite dimensional manifold $H(S)$ in a way that it is not possible to construct a sequence of functions $v_i(t) \in H(S)$ that integrate to 1, are nonnegative, and converge pointwise to a Dirac delta function. Along this line of thinking, Robertson [18], Rao [17], and Wegman [26-28] restricted $H(S)$ in a way that ensured the existence of the maximum likelihood estimate. The restriction considered by the above essentially requires that $H(S)$ contains only unimodal density functions that are monotone on either side of the mode. Starting with the assumption of known mode, it can be shown that the ML estimate is a piecewise constant density function. Subsequently, Wegman [26-28] proposed and analyzed an adaptive procedure that also seeks the mode, hence the method works for unknown mode.

There is no smoothing involved in the above procedure, so a practical disadvantage is the lack of smoothness of the resulting estimates.

A second method for choosing a convenient $H(S)$ is to consider the class of monotone failure rate distribution functions. This class is motivated by concepts in reliability. Such an approach has been followed by Marshall and Proschan [14], Crow and Shimi [4], Rao [17]. A probability distribution function $F(x)$ is said to be IFR (increasing failure rate) on $[0,T]$ if $F(0) = 0$ and $-\log[1 - F(x)]$ is convex on $[0,T]$. Under this constraint, the maximum likelihood estimate of the distribution is well behaved. Crow and Shimi [4] derive its explicit structure and prove that it converges uniformly almost surely to the true F as the number of data points goes to ∞.

A third and quite different approach was originated by Good and Gaskins [10]. Their thinking was that since the original maximum likelihood problem usually has as a solution a sum of Dirac delta functions, we should modify the likelihood function in a way that will penalize such nonsmooth solutions. Essentially we have a constrained optimization problem, namely we seek the maximization of $L(v)$ under the constraint that v is "sufficiently smooth." Let $\Phi(v)$ be a functional $\Phi: H(S) \to R$, which acts as a measure of the smoothness of v, i.e., the smoother v is, the smaller Φ becomes.

Following the philosophy of constrained optimization theory, Good and Gaskins proposed the Φ-penalized likelihood of $v \in H(S)$ as:

$$L_p(v) = \prod_{i=1}^{n} v(x_i)\exp(-\Phi(v)) \tag{2}$$

They proposed the Φ-penalized maximum likelihood estimate of v (MPLE) as the solution to the optimization problem:

$$\left.\begin{array}{l} \text{maximize } \; L_p(v) \; \text{ subject to} \\ v \in H(S), \; \int v(t)\,dt = 1, \; v(t) \geq 0, \; \forall\, t \in S \end{array}\right\} \tag{3}$$

One choice for $\Phi(v)$ that Good and Gaskins considered is

$$\Phi(v) = \int_S [v'(t)]^2[v(t)]^{-1}\,dt \tag{4}$$

Recently, de Montricher, Tapia and Thompson [6] studied and established the existence and uniqueness of the MPLE for wide choices of Φ. They also found that one of Good and Gaskins estimates is a positive exponential spline with knots at the sample points. This is an interesting ramification, justifying in a new way the estimation of probability density functions using splines studied by G. Wahba [25].

However, many important statistical considerations regarding consistency, unbiasedness, convergence rate, choice of penalty term and numerical implementation are still unanswered and await further investigation. In a recent Ph.D. thesis, Scott [20] considered among other things a discretized form of the MPLE for numerical implementation.

The most general convergence result for the MPLE proven by Good and Gaskins [10] is the following.

For any $v \neq v_o$, v_o = true density, $E[L_p(v_o)] > E[L_p(v)]$ for n large enough. This result does justify the whole procedure of MPLE, but does not imply consistency.

IV. KERNEL DENSITY ESTIMATES

Let $x_1,\ldots,x_n$ be a sequence of statistically independent random variables distributed according to a common probability density function $f(x)$. A kernel density estimate $\hat{f}_n(x)$ is defined as

$$\hat{f}_n(x) = \int_{-\infty}^{+\infty} K_n(x,y)dF_n(y) = n^{-1}\sum_{i=1}^{n} K_n(x,x_i) \tag{5}$$

where $F_n(y)$ is the empirical distribution function based on n observations and has the staircase form. The above integral shows the philosophy of using a kernel function $K_n(x,y)$ to smear or smooth the empirical estimate.

The choice of $K_n(x,y)$ is therefore very important for the performance of the resulting estimator. Probability density estimation is similar in many aspects to spectral density estimation. One concept common to both areas is smoothing. Data smoothing windows have been used extensively in estimating the spectrum of time series. The analogy between the two problems has been discussed by Bartlett [1]. Recently Papoulis [15] derived the optimal smoothing window for spectral estimation.

Tracing back to kernel type density estimators, Rosenblatt [19] considered an estimate of the form:

$$\hat{f}_n(x) = (2h)^{-1}[F_n(x + h) - F_n(x - h)] \tag{6}$$

where $F_n(x)$ is, as before, the empirical distribution of the staircase type.

The estimate (6) is a special case of (5) for

$$K_n(x,y) = \begin{cases} (2h)^{-1} & \text{for } |x - y| \le h \\ 0 & \text{otherwise} \end{cases}$$

(6) can be considered as a rudimentary smoothing of the empirical distribution F_n.

h controls the width of the smoothing interval. As the number n of observations grows, the smoothing interval should shrink because the local statistical structure can be identified more finely. This heuristic argument turns out to be rigorously justified in kernel density estimation.

Rosenblatt showed that the mean square error achievable by the estimator (6) is:

$$E[\hat{f}_n(x) - f(x)]^2 = (2hn)^{-1} \cdot f(x) + 36^{-1}h^4|f''(x)|^2 + 0(h^4 + (hn)^{-1}) \tag{7}$$

The assumption he used was the existence of the first three derivatives of $f(x)$ at x. If h is chosen as $h = Kn^{-\alpha}$, $\alpha > 0$, it was found that the optimum choice for α is 1/5 and the optimum K is

$$K_o = \{\frac{9}{2} \cdot f(x) \cdot [f''(x)]^2\}^{1/5} \tag{8}$$

But if instead of the mean square error we use the mean integrated square error

$$E\int_{-\infty}^{+\infty} [\hat{f}_n(x) - f(x)]^2 dx$$

as performance criterion, then the optimum α remains 1/5 while the optimum K becomes

$$K_1 = \{\frac{2}{9}\int_{-\infty}^{+\infty} [f''(x)]^2 dx\}^{-5/2} \tag{9}$$

In practice we are estimating $f(x)$ and it is unrealistic to attempt to use the optimal values of K, namely K_1 or K_o. Hence K_1 and K_o have only theoretical importance. The method is still useful, however, because for any positive K and $\alpha = 1/5$ we have pointwise and integrated mean square consistency. Rosenblatt also proposed kernel estimates with

$$K_n(x,y) = h^{-1}K[(x - y)h^{-1}] \tag{10}$$

and where K is a nonnegative even around zero function such that

$$\int_{-\infty}^{+\infty} K(s)\ ds = 1 \tag{11}$$

$$\int_{-\infty}^{+\infty} K^2(y)\ dy < \infty \tag{12}$$

$$\int_{-\infty}^{+\infty} K(y)\,|y|^3\ dy < \infty \tag{13}$$

If h is chosen optimally and f has derivatives of the first three orders, then the mean square error is no smaller than $0(n^{-4/5})$. Again, the optimal choice of h involves knowledge of f and is impractical to achieve. Parzen [16] considered kernel estimates with the requirement that $K(y)$ as defined above is nonnegative, even around zero and

$$\sup_{-\infty<y<\infty} K(y) < \infty \tag{14}$$

$$\int_{-\infty}^{+\infty} K(y)\ dy < \infty \tag{15}$$

$$\lim_{|x|\to\infty} |x|K(x) = 0 \tag{16}$$

He named $K(y)$ weighting function and showed that the asymptotic variance of $\hat{f}_n(x)$ is

$$(nh_n)^{-1} \cdot f(x) \cdot \int_{-\infty}^{+\infty} K^2(y)\, dy$$

at the continuity points of f. Hence for the variance to go to zero we must require $nh_n \to \infty$. Also, for the bias $E[\hat{f}_n(x) - f(x)]$ to go to zero at the continuity points of $f(x)$, we must have $h_n \to 0$. With some more conditions, Parzen shows that $\hat{f}_n(x)$ is asymptotically normal.

In a recent paper [5], Davies and Wegman developed a sequential nonparametric density estimation procedure using kernel estimates of the Parzen type. Their procedure develops a stopping rule with regard to the number of blocks of samples used and is sequential in the above sense. However, it is not sequential computationally, i.e., when more data are acquired the old estimate is thrown away and the new one is constructed from scratch. This is due to the need of updating h_n. They consider consecutive blocks of data of size m and they form the difference

$$V_n(x) = \hat{f}_{nm}(x) - \hat{f}_{(n-1)m}(x) \tag{17}$$

where n is the block index. The stopping rule is activated when the norm of $V_n(x)$ falls below a threshold ε.

Until now, the kernel function K has been left unspecified with the only requirement being the satisfaction of the previous constraints.

In a recent work, Bennett, de Figueiredo and Thompson [2] considered the use of a B-spline as kernel function K.

Epanechnikov [7] studied the kernel function optimization problem. He restricted the class of kernel functions by imposing an arbitrary constraint of unit variance, i.e.,

$$\int_{-\infty}^{+\infty} y^2 K(y)\, dy = 1 \tag{18}$$

He then uses an approximation to the integrated mean square error between $\hat{f}_n(x)$ and $f(x)$. Among the class of unit variance kernel functions he finds the one that minimizes the approximate integrated mean square error. The unit variance constraint facilitates his solution, but makes the result suboptimal, since no such constraint is required by the theorems on consistency. Any constraint on the $2m^{th}$ moment of the kernel function $(m = 1,2,\ldots,)$ can be similarly imposed, producing a different suboptimal solution K_m.

We will show how we can avoid all of the disadvantages of the suboptimal solution of Epanechnikov and still obtain a simple and appealing optimal kernel function.

In a series of papers, G Wahba [23 - 25] recently compared the convergence properties of the following density estimators:

a) The ordinary histogram.

b) Certain orthogonal series estimates.

c) Kernel-type estimates.

d) Polynomial interpolation algorithm.

e) Interpolating spline methods with fixed and variable knots.

We will summarize her results after a few definitions are given. Let

$$\|f^{(m)}\|_p = \begin{cases} \left| \int_S [f^{(m)}(t)]^p \, dt \right|^{1/p} , & 1 \le p < \infty \\ \sup_{t \in S} |f^{(m)}(t)| , & p = +\infty \end{cases} \tag{19}$$

$f^{(i)}$ denotes the ith derivative of f. Let $W_p^{(m)}(U)$ be the Sobolev space of functions defined on S whose $m - 1$ first derivatives are absolutely continuous, whose m^{th} derivative is in $L_p(S)$ and $\|f^{(m)}\|_p \le U$. The parameters (m,p,U) quantify the degree of smoothness of the elements $f \in W_p^{(m)}(U)$. Let

$$s(m,p) = (2m - 2/p)/(2m + 1 - 2/p) \tag{20}$$

In what follows as measure of closeness of an estimate $\hat{f}_n(x)$ to the true

$f(x)$ we will use the pointwise mean square error

$$E[\hat{f}_n(x) - f(x)]^2$$

Farrel [8] showed that for any estimate $\hat{f}_n(x)$ and $f \in W_p^{(m)}(U)$, the mean square error cannot have a rate of convergence better than $n^{-s(m,p)}$. In [23-25], Wahba proved that the density estimators a) - e) achieve the above convergence rate in the sense that

$$E[\hat{f}_n(x) - f(x)]^2 \leq D_j n^{-s(m,p)} \tag{21}$$

$$j = a, b, c, d, e$$

where $\{D_j, j = (a), (b), (c), (d), (e)\}$ are constants corresponding to each of the above estimates.

We will concentrate only on the case (c) because we are concerned with kernel estimates. We assume known a uniform upper bound to the class of unknown densities, i.e.,

$$\sup_x |f(x)| \leq V \tag{22}$$

We now assume that the true density function f is smooth enough to belong to $W_p^{(m)}(U)$

$$f \in W_p^{(m)}(U) \qquad m \geq 2$$

Let $Q(y)$ be a real valued kernel function on $(-\infty,+\infty)$, satisfying the following conditions:

$$\text{i)} \quad \sup_{-\infty<y<+\infty} |Q(y)| < \infty, \qquad \text{ii)} \quad \int_{-\infty}^{+\infty} |Q(y)|\,dy < \infty$$

$$\text{iii)} \quad \lim_{|y|\to\infty} |yQ(y)| = 0, \qquad \text{iv)} \quad \int_{-\infty}^{+\infty} Q(y)\,dy = 1 \tag{23}$$

$$\text{v)} \quad \int_{-\infty}^{+\infty} y^s Q(y)\,dy = 0 \quad \text{for} \quad s = 1,2,\ldots,m-1$$

vi) $\int_{-\infty}^{+\infty} |y|^{m-1/p} |Q(y)| dy < \infty$

The kernel density estimate $\hat{f}_n(x)$ using Q(y) as a kernel function is

$$\hat{f}_n(y) = (nh_n)^{-1} \sum_{k=1}^{n} Q[h_n^{-1}(y - x_k)]$$

where h_n must be chosen so that

$$h_n \to 0 \ , \ nh_n \to +\infty \quad \text{as} \quad n \to \infty$$

We define the constants

$$B = V \int_{-\infty}^{+\infty} Q^2(y) dy$$

$$\alpha = m - p^{-1} \quad , \quad q = (1 - p^{-1})^{-1} \tag{24}$$

$$A = [\int_{-\infty}^{+\infty} |Q(y)| |y|^{\alpha} dy]^2 \cdot [(m - 1)!]^2 \cdot [(m - 1)q + 1]^{-2/q}$$

$(\alpha > 0)$

G. Wahba [24] developed the following upper bound to the pointwise mean square error:

$$E[\hat{f}_n(y) - f(y)]^2 \leq V^2 A h_n^{2\alpha} + B(nh_n)^{-1} \tag{25}$$

The above bound is very useful because it converges to zero as $n \to \infty$ and it depends only on (Q,U,V,m,p). Let $K_n = nh_n$. The choice of K_n that minimizes the given upper bound is

$$K_n = [\alpha^{-1} A^{-1} V^{-2} \cdot B]^{1/(2\alpha+1)} \cdot n^{s(m,q)} \tag{26}$$

It is practically feasible to design K_n optimally because it requires only minimal knowledge of the density. Let

$$C = (1 + 2\alpha)(2\alpha)^{-2\alpha} U^{2/(1+2\alpha)} V^{2\alpha/(1+2\alpha)}$$

$$\cdot [(m - 1)![q(m - 1) + 1]^{1/q}]^{-2/(1+2\alpha)} \tag{27}$$

$$G(Q) = \int_{-\infty}^{+\infty} |Q(y)||y|^{\alpha}dy \cdot [\int_{-\infty}^{+\infty} Q^2(y)dy]^{\alpha} \tag{28}$$

For the optimal choice of K_n , the corresponding mean square error bound is:

$$E[\hat{f}_n(y) - f(y)]^2 \leq CG(Q) \cdot n^{-s(m,q)}[1 + 0(1)] \tag{29}$$

where $0(1) \to 0$ as $n \to \infty$. The choice of Q that will minimize G under the constraints (i) - (vi) is the optimal in terms of minimizing the upper bound given. The amount of data smoothing increases with the size of the support $[-T,T]$ of the kernel function Q . Too much smoothing would smear out the fine statistical structure of the data and it would impose an unduly high computational load in the implementation. Thus a finite T should be used. If $Q(y)$ is required to have a finite support and be bounded, the constraints (i), (ii), (iii), and (vi) are satisfied. Only (v) is difficult to satisfy, the difficulty increasing with the degree of smoothness m . If we let $m = 2$ and constrain Q to be nonnegative and even, i.e., $Q(y) = Q(-y)$, $Q(y) \geq 0$, (v) is satisfied and we have minimal smoothness requirements on the class of density functions. Also the nonnegativity of Q is beneficial for the small sample performance of the density estimate. By a simple change of variables we transform the interval $[0,T]$ into $[0,1]$. Let

$$b = (2T)^{-1} \quad , \quad K(z) = Q(zT) \quad \text{for} \quad z \in [-1,1] \tag{30}$$

To comply with constraint (i), we require that $K(z) \leq U_o < +\infty$. In order that we do not restrict extremely the space of eligible K's , we assume that U_o is a large positive constant. Let

$$H(K) = T^{-2\alpha-1}G(K) \tag{31}$$

Due to the eveness of $K(z)$, the minimization of the bound takes the following structure.

Minimize

$$H(K) = \int_0^1 K(z)z^{\alpha}dz[\int_0^1 K^2(z)dz]^{\alpha}$$

under the constraints

$$\begin{cases} 0 \leq K(z) \leq U_o \;, \; z \in [0,1] \\ \int_0^1 K(z)dz = b \end{cases} \tag{32}$$

The strategy we will follow is to fix the value of the integral $\int_0^1 K^2(z)dz = q$, minimize $H(K)$ for a fixed q and then vary q to find the overall minimum. The problem now is: Minimize

$$F(K) = \int_0^1 K(z)\, z^{\alpha}dz = H(K) \cdot q^{-\alpha} \tag{33}$$

under the constraints

$$0 \leq K(z) \leq U_o \qquad (U_o > b)$$

$$\int_0^1 K(z)dz = b \tag{34}$$

$$\int_0^1 K^2(z)dz = q$$

Using the Hölder inequality we can easily show [12] that all the values q can take are in the interval $[b^2, bU_o]$. Also, for any $q \in [b^2, bU_o]$ there exists a rectangular function satisfying the constraints (34). We further note that z^{α} is monotone increasing in $[0,1]$ for $1 < \alpha < 2$. Therefore, the minimizing $K(z)$ under the constraints (c) must be monotone decreasing. The nonnegativity constraint combined with the monotonicity requirement dictates that the optimal $K(z)$ must have one of two possible forms:

(e1) $K(z)$ = positive for all $z \in [0,1)$

(e2) $K(z)$ = positive for all $z \in [0,u)$

and zero for $z \in [u,1]$

We will use the method of Lagrange multipliers in order to solve the constrained minimization problem. Let A_1, B_1 be Lagrange multipliers. Define, for $0 < u \leq 1$ the functional:

$$L_u(K, A_1, B_1) = \int_0^u K(z)z^{\alpha}dz - A_1 \int_0^u K(z)dz + 2^{-1}B_1 \int_0^u K^2(z)dz$$

$$= -2^{-1}B_1^{-1} \int_0^u (A_1 - z^{\alpha})^2 dz + 2^{-1}B_1 I(K) \tag{35}$$

where

$$I(K) = \int_0^u [K(z) - B_1^{-1}(A_1 - z^{\alpha})]^2 dz \tag{36}$$

If $B_1 < 0$, minimization of L_u is equivalent to the maximization of $I(K)$. This would require a monotone increasing $K(z)$ which is unacceptable. Hence we must have $B_1 > 0$. For $B_1 > 0$, $A_1 < 0$ the minimizing solution is $K(z) = 0$, unacceptable. Hence we must have $A_1 > 0$, $B_1 > 0$. Then the minimizing solution is:

$$K_o(z) = B_1^{-1}(A_1 - z^{\alpha}) \quad , \quad z \in [0, u]$$

A_1 and B_1 must be found from the constraints. We must do this separately for solutions of type (e1) and (e2).

For solutions of type (e1), we must have $A_1 \geq 1$, $u = 1$, $B_1 > 0$. Substituting $K_o(z)$ into (34) and solving for A_1, B_1 in terms of (α, q, b) we find:

$$\left.\begin{aligned} B_1 &= [(1 + 2\alpha)^{-1} - (1 + \alpha)^{-2}]^{\frac{1}{2}}(q - b^2)^{-\frac{1}{2}} \\ A_1 &= (1 + \alpha)^{-1} + b(q - b^2)^{-\frac{1}{2}}[(1 + 2\alpha)^{-1} - (1 + \alpha)^{-2}]^{\frac{1}{2}} \end{aligned}\right\} \tag{37}$$

The conditions $A_1 \geq 1$, $q \geq b^2$ imply:

$$1 \leq qb^{-2} \leq 1 + (1 + 2\alpha)^{-1} \tag{38}$$

Hence, in the parameter region specified by (38), we have a solution of type (e1).

For minimizing solutions of type (e2), we must have

$$\left.\begin{aligned} &\int_0^u K_o(z)dz = b \quad , \quad A_1 = u^\alpha \\ &\int_0^u K_o^2(z)dz = q \quad , \quad 0 < u < 1 \end{aligned}\right\} \tag{39}$$

Substituting the function $K_o(z)$ into (39), we can solve for A_1, B_1, u with result

$$\begin{aligned} u &= b^2 q^{-1}[1 + (1 + 2\alpha)^{-1}] \\ B_1 &= \alpha b^{-1}(1 + \alpha)^{-1}u^{\alpha + 1} \\ A_1 &= u^\alpha \end{aligned} \tag{40}$$

The condition $u < 1$ defines as parameter region where (e2) is the minimizing solution, the region

$$qb^{-2} \geq 1 + (1 + 2\alpha)^{-1} \tag{41}$$

and the solution is complete. If we denote by $H_o(q)$ the minimum achievable value of $H(K)$ for a fixed q, we readily find:

$$b^{-1-2\alpha}H_o(q) = \begin{cases} (1 + \alpha)^{-1}[1 - \alpha(1 + 2\alpha)^{-\frac{1}{2}}(qb^{-2} - 1)^{\frac{1}{2}}](qb^{-2})^\alpha \\ \qquad \text{for} \quad 1 \leq qb^{-2} \leq 1 + (1 + 2\alpha)^{-1} \\ (1 + 2\alpha)^{-1}[1 + (1 + 2\alpha)^{-1}]^\alpha \\ \qquad \text{for} \quad qb^{-2} \geq 1 + (1 + 2\alpha)^{-1} \end{cases} \tag{42}$$

The upper expression corresponds to a solution of type (e1) and the lower to one type of (e2). It is straightforward to show that

$$H_o'(q) < 0 \quad \text{for} \quad 1 \leq qb^{-2} \leq 1 + (1 + 2\alpha)^{-1}$$

Therefore, $H_o(q)$ is monotone decreasing in the above region and constant for $qb^{-2} \geq 1 + (1 + 2\alpha)^{-1}$. As a consequence, among all solutions (e1) the best is the one corresponding to $qb^{-2} = 1 + (1 + 2\alpha)^{-1}$. As for the solutions of type (e2), all of them achieve the same H_o. However, among them the one with $qb^{-2} = 1 + (1 + 2\alpha)^{-1}$ has the best data smoothing effect and is intuitively preferable for small sample performance.

As a conclusion, we see that the kernel function

$$K_o(z) = b(1 + \alpha^{-1})(1 - |z|^{\alpha}) \quad , \quad |z| \leq 1 \tag{43}$$

minimizes the given upper bound to the pointwise mean square error to the kernel density estimate. The resulting H is:

$$H(K_o) = b^{1+2\alpha}(1 + 2\alpha)^{-1}[1 + (1 + 2\alpha)^{-1}]^{\alpha} \tag{44}$$

The optimal G, denoted as G^*, is:

$$G^* = H(K_o) \cdot T^{2\alpha+1} = (2 + 4\alpha)^{-1}[4^{-1} + (4 + 8\alpha)^{-1}]^{\alpha} \tag{45}$$

Interestingly, G^* does not depend on T. This provides us with the freedom to choose T on the basis of small sample performance considerations and the degree of smoothing desired. Hence, its choice must be made on empirical grounds.

Finally, going back to the interval [-T,T] we find that the optimal kernel function is

$$Q(y) = (1 + \alpha^{-1})(2T)^{-1}[1 - T^{-\alpha}|y|^{\alpha}] \tag{46}$$

for $|y| \leq T$ and $\alpha = 2 - p^{-1}$.

The simple form of the found optimal kernel renders it useful for implementation. Generalization of the result to vector random variables and multivariate densities is not feasible yet because it requires extension of Wahba's result [24]. It seems that such an extension is feasible because Farrell's [8] asymptotic achievable rate is valid for multivariate densities.

REFERENCES

1. Bartlett, M.S. (1963) "Statistical estimation of density functions," Sankhyā (A) 25: 245-254.

2. Bennett, J.O, R.J.P. de Figueiredo, J.R. Thompson (1974) "Classification by means of B-spline potential functions with applications to remote sensing," Proc. Sixth Southeastern Symp. on System Theory, Baton Rouge, La.

3. Cover, T. (1972) "A hierarchy of probability density function estimates," in Frontiers of Pattern Recognition (S. Watanabe, ed.), Academic Press, New York, pp. 83-98.

4. Crow, L.H. and I.N. Shimi (1972) "Maximum likelihood estimation of life-distributions from renewal testing," Ann. Math. Stat. 43: 1827-1838.

5. Davies, H.J. and E.J. Wegman (1975) "Sequential nonparametric density estimation," IEEE Trans. Inf. Th. IT-21: 619-628.

6. de Montricher, G.F., R.A. Tapia, and J.R. Thompson (1975) "Nonparametric maximum likelihood estimation of probability densities by penalty function methods," Ann. of Stat. 3: 1329-1348.

7. Epanechnikov, V.A. (1969) "Nonparametric estimates of a multivariate probability density," Theor. Prob. Appl. 14: 153-158.

8. Farrell, R.H. (1972) "On best obtainable asymptotic rates of convergence in estimation of a density function at a point," Ann. Math. Stat. 43: 170-180.

9. Fisher, R.A. (1922) "On the mathematical foundations of theoretical statistics," Phil. Trans. Roy. Soc. London Ser. A. 222: 309-368.

10. Good, I.J. and R.A. Gaskins (1971) "Nonparametric roughness penalties for probability densities," Biometrika 58: 255-277.

11. Jenkins, G.M. and D.G. Watts (1968) Spectral Analysis and Its Applications. Holden-Day, San Francisco.

12. Kazakos, D. "Optimal choice of the kernel function for the Pazzen kernel-type density estimators," to appear.

13. Lucky, R.W. (1973) "A survey of the communication theory literature: 1968-1973," IEEE Trans. Inf. Th. IT-19: 725-739.

14. Marshall, A. and F. Proschan (1965) "Maximum likelihood estimation for distributions with monotone failure rate," Ann. Math. Stat. 36: 69-77.

15. Papoulis, A. (1973) "Minimum bias windows for high-resolution spectral estimates," IEEE Trans. Inf. Th. IT-19: 9-12.

16. Parzen, E. (1962) "On estimation of a probability density function and mode," Ann. Math. Stat. 33: 1065-1076.

17. Rao, B.L.S.P. (1970) "Estimation for distributions with monotone failure rate," Ann. Math. Stat. 41: 507-519.

18. Robertson, T. (1967) "On estimating a density which is measurable with respect to a σ-lattice," Ann. Math. Stat. 38: 482-493.

19. Rosenblatt, M. (1956) "Remarks on some nonparametric estimates of a density function," Ann. Math. Stat. 27: 832-837.

20. Scott, D.W. (1976) Nonparametric Probability Density Estimation by Optimization Theoretic Techniques, Ph.D. Thesis, Department of Mathematical Sciences, Rice University, Houston, Texas.

21. Wagner, T.J. (1973) "Strong consistency of a nonparametric estimate of a density function," IEEE Trans. Syst., Man, Cyber. SMC-3: 289-290.

22. Wagner, T.J. (1975) "Nonparametric estimates of probability densities," IEEE Trans. Inf. Th. IT-21: 438-440.

23. Wahba, G. (1973) "Interpolating spline methods for density estimation. II. Variable knots," Tech. Report #337, Department of Statistics, Univ. of Wisconsin, Madison, Wisc.

24. Wahba, G. (1975) "Optimal convergence properties of variable knot, kernel, and orthogonal series methods for density estimation," Ann. Stat. 3: 15-29.

25. Wahba, G. (1975) "Interpolating spline methods for density estimation. I. Equi-spaced knots," Ann. Stat. 3: 30-48.

26. Wegman, E.J. (1970) "Maximum likelihood estimation of a unimodal density function," Ann. Math. Stat. 41: 457-471.

27. Wegman, E.J. (1970) "Maximum likelihood estimation of a unimodal density, II," Ann. Math. Stat. 41: 2169-2174.

28. Wegman, E.J. (1971) "A note on the estimation of the mode," Ann. Math. Stat. 42: 1909-1915.

29. Wegman, E.J. (1972) "Nonparametric probability density estimation: I. A summary of available methods," Technometrics 14: 533-546.

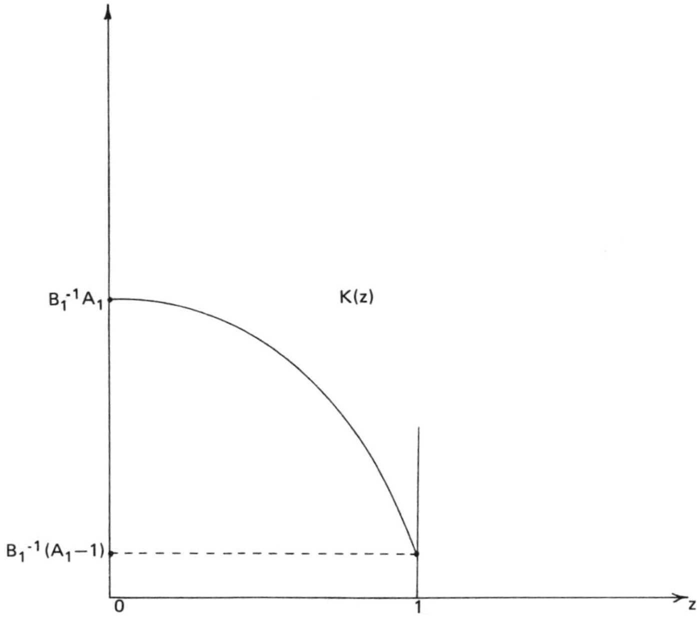

Figure 1 Solution of form e1.

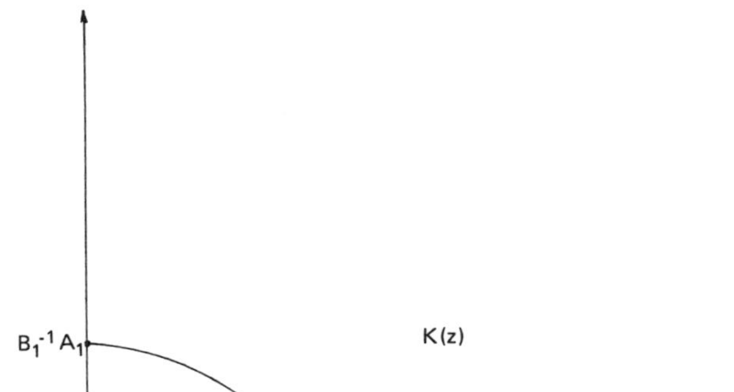

Figure 2 Solution of form e2.

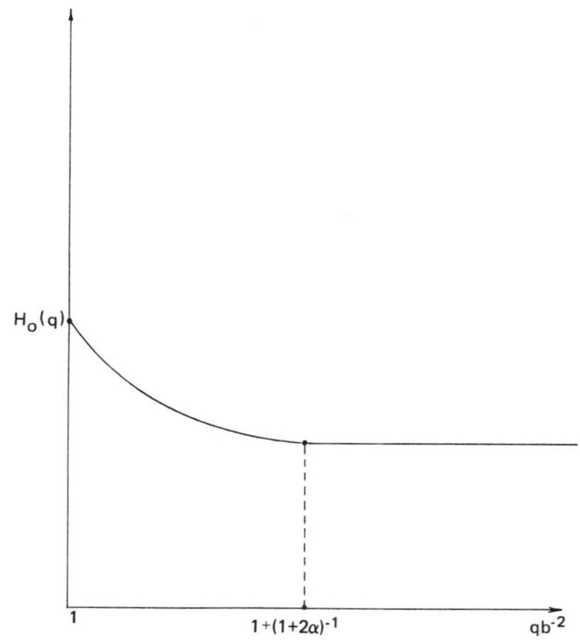

Figure 3 Optimal value of the bound as a function of q.

INDEX